BIRDS BY NIGHT

For
Laura *and* Rowan

BIRDS BY NIGHT

by Graham Martin

Illustrated by
JOHN BUSBY

T & A D POYSER
London

© Graham Martin 1990

ISBN 0 85661 059 3

First published in 1990 T & A D Poyser Ltd
24–28 Oval Road, London NW1 7DX

Text set in Garamond
Typeset by Paston Press, Loddon, Norfolk
Printed and bound in Great Britain by
Mackays of Chatham PLC, Chatham, Kent

British Library Cataloguing in Publication Data
Martin, Graham
    Birds by night.
    1. Birds. Influence of night
    I. Title
    598.251

ISBN 0–85661–059–3

Contents

List of Illustrations

List of Figures

List of Tables

Preface

My original motivation for writing this book was to explore a single question: "How is that owls are nocturnal?" Like many simple questions, innocently put, there is no simple answer which will satisfy the curious. Owls cannot be examined in isolation and the more general question of nocturnal activity in birds must be examined if the question of nocturnality in owls is to be viewed in its correct context. The answer to the question is a complicated one which in its telling opens up subject areas and leads down avenues of thought which the questioner perhaps never suspected would be relevant to the initial enquiry.

To address the question of nocturnality in birds in anything more than a superficial way requires knowledge drawn from diverse areas, often across defined disciplinary boundaries. Help must be sought from field ornithologists, sensory scientists, ecologists and physicists. Information must be gathered on the natural history and behaviour of many nocturnal bird species, on the occasional nocturnal behaviours of otherwise strictly diurnal birds, and also on owl species which are mainly diurnal and crepuscular in their habits, as well as those which are strictly nocturnal. Information must also be gathered on the sensory capacities of birds and these must be compared with those of other animal species including ourselves. Finally, there must be information on the sensory problems which night- and day-time environments actually present to an animal.

This book attempts to weave such information together and in so doing capture something of the complexity of the relationships which exist between any animal and its environment. Comparing and contrasting information is particularly important in these arguments for it is only when the sensory capacities and natural history of the nocturnal birds are compared with those of diurnal species, that the features which make strictly nocturnal birds so special can be appreciated.

Many of the ideas presented here have germinated over a long period and were formed as the result of discussions and correspondence with a wide range of people. It would be impossible to acknowledge here all who contributed in divers ways to the ideas of this book. A special acknowledgement must go to Dr Ian Gordon who started me out on owls in 1970. Listed in alphabetical order, the following people have all provided crucial encouragement to me and/or information on various topics related to the question of nocturnality in birds: Dr R. R. Baker, Professor P. Berthold, Dr J. K. Bowmaker, Dr M. Brooke, Dr J. Cohen, Professor P. R. Evans, Dr G. Goss-

Custard, Dr G. Hirons, Professor A. S. King, Professor M. F. Land, C. Mead, Professor W. R. A. Muntz, Dr D. Pearson, Dr P. Soilleux, Professor K. H. Voous and Dr S. R. Young. The wardens and their helpers at the following Bird Observatories kindly provided data on the occurrence of nocturnal migratory activity in birds: Neville McKee (Copeland), David Walker (Calf of Man), Tim Dean (Walney), Peter Hope Jones (Bardsey), Sean McMinn (Dungeness), John Cudworth (Spurn), Peter Howlett (Fair Isle).

Finally, two special acknowledgements: to my wife, Marie-Anne, for her help and encouragement which made sure that this book was written; and to Wol, my Tawny Owl companion for twenty years. He is still a source of wonder and inspiration, but is still totally unimpressed by his own ability to be active at night.

CHAPTER 1

The dead of night?

Of all clichés, "in the dead of night", must be one of the most overworked but also one of the most evocative and universally understood. It conjures up thoughts of intrigue, drama and mystery by combining images of total darkness with ideas of complete stillness and death. A frightening and morbid picture that can immediately set the scene for the narrative to follow. But is it accurate? Is it just a literary device which plays upon our ignorance of what night-time truly consists of?

From the perspective of the natural world that cliché is far from accurate, for the night is never still or dead, nor is it totally dark. The night is as alive and ever changing as the day-time. Night-time is certainly mysterious but only because of our ignorance of it both in terms of common experience and of scientific knowledge. It is essential to appreciate that the night is not just "day-time without the sun", it is a different world populated by different animals or animals completing tasks in different sorts of ways, guided by different cues to those which would be relied upon during daylight. For truly nocturnal

1

animals their natural world is perhaps no more exacting than the day-time world is for diurnally active creatures, there are simply different problems which are solved in different ways.

More images of mystery and intrigue have been woven around the nocturnally active birds than probably any other group of animals. It is the owls in particular which are regarded as the living embodiment of all that the phrase "in the dead of the night" implies. A dramatist has only to mention the hoot of an owl and the scene is set for mystery.

It is the aim of this book to remove some of the ignorance upon which these literary images are based; to provide a broad perspective with which to view the nocturnal activities of many animals and in particular to address some of the mystery which surrounds the nocturnal activities of birds. This is not an attempt simply to debunk a few myths but rather to replace some assumptions with more solid factual and theoretical ideas. When the reader reaches the end of this book it is hoped that the "dead of night" and all that goes on in it, will have taken on new meaning and images, but that these will be no less interesting than those which they have replaced.

THE NOCTURNAL HABIT IN BIRDS

Activity at night is rare amongst birds and a strictly nocturnal habit, in which all aspects of the life cycle are completed between dusk and dawn, is found in probably less than 3% of the world's species. The opposite habit of diurnality (ceasing activity when light levels decrease to those experienced around dusk and staying inactive until around dawn) is so common among birds that it is often taken for granted. The corollary has been that nocturnality in birds is regarded as so unusual that nocturnal species have been frequently regarded as anomalous or even curiosities of the bird world. One result of this has been the tendency in both popular imagination, religious symbolism and in literature from many cultures, to endow nocturnal birds, especially the owls, with special powers. See, for example, the discussions in Armstrong (1958), Sparks and Soper (1970), Bunn et al (1982) and Burton (1984), which describe the many symbolic and other cultural functions of owls in folklore, religion and literature.

This tendency to endow owls with special powers has extended into popular ornithology and has even spilled over into the scientific field where rather simple explanations of the nocturnal habit, invoking special or exceptional sensory powers, have been readily proposed and perhaps too readily or uncritically accepted.

Such treatment is not of course unique to discussions of nocturnality and a parallel to this can be found in the early scientific interpretations and folklore beliefs that surrounded the phenomenon of bird migration. That long distance migratory birds possessed a "sixth sense" which could account for their navigational abilities was an idea which held sway for many decades. Scientific searches for the sixth sense have revealed, however, a complex of sensory abilities and behavioural strategies which birds may employ in various ways during the migratory phases of their lives [see, for example, the reviews of Able (1980), Baker (1984), Presti (1985) and Wiltschko and Wiltschko (1988a,b)]. In short, there is not a simple answer to bird migration and navigation.

The argument which underpins this book on nocturnality also parallels that regarding the modern understanding of bird migration. It is argued that strict nocturnality in birds is seen to be achieved not by reliance upon one exceptional sensory capacity but by the integrated employment of various sensory cues which must be correctly interpreted and which work in concert with various behavioural adaptations. It is the unravelling of the sensory capacities and the behavioural adaptations that accompany the nocturnal habit which is the burden of this book. It will be seen, however, that there are many unanswered questions.

DIURNAL MAN AND NIGHT-TIME

It is worth considering briefly why nocturnality in birds should seem so strange and thus the ready source of cultural and scientific myths. One reason for this probably lies in man's own mainly diurnal habits. In modern western culture, at least, the reliance upon so-called "artificial light sources" has led to a failure to understand at an everyday sort of level the problems posed by the night-time world. On the whole, humans are out and about during the day and go to sleep at night. Therefore, from an anthropocentric view point the diurnal birds seem to do the "sensible" or "obvious" thing. If one wants to be active and abroad then the best time to do so is during daylight – going about one's business at night is seen at the least to be deliberately making life difficult for oneself, and at the worst as revealing highly suspicious motives for one's actions. However, when viewed against the background of other mammals our own diurnal habits are somewhat unusual, for on the whole mammals are nocturnal in their activities.

This nocturnal–diurnal difference in the activity patterns of the majority of mammals and birds has led the field mammologist and the field ornithologist to develop quite different skills. Consideration of these skills serves to demonstrate how the dirunally active animal seems to rely primarily upon vision to gain information instantaneously about objects both close to and distant. The nocturnally active mammals, however, seem to rely upon other senses which may not necessarily yield such detailed instantaneous infor-mation but can often provide reliable information about past events of importance to the animal.

Even with the aid of modern equipment such as radio trackers and image amplifiers, field mammologists still need to develop expertise in detecting and interpreting the signs of where their animal *has* been. From these observations experienced mammologists can often reconstruct the activity patterns of their quarry, hours or even days before their observations were made. The field ornithologist on the other hand needs to develop quite different skills of rapid visual identification and an ability to interpret the briefest sightings of a bird

which typically leaves behind it few signs of its activity that can be analysed at leisure. On the whole ornithologists and bird watchers are not satisfied to see a tree where a bird sat the day before, while mammologists are often delighted to find, for example, scratch marks or faeces produced by their quarry at the base of a tree perhaps days previously. In short, diurnal bird watchers and dirunal birds would seem to guide most of their behaviour by information instantaneously released by the animal of interest, while the nocturnal mammologists and their mammals are guided by less instantaneous, but often persistent information left behind by the animal of their interest [see, for example, the detailed body of work on the nocturnal activities of the red fox, *Vulpes vulpes* (Macdonald, 1987)].

For humans the night environment seems to present an immediate barrier to activity. Most people regard themselves as inadequately equipped to be abroad at night and certainly many activities which we take for granted during the day cannot be performed once the sun has set. We can gain some comfort from our own apparent inadequacy at night compared with other common, but nocturnal, mammals by reference to the fact that these mammals may have at their disposal sensory capacities which humans lack. In some ways these capacities provide the easy solution to the question, "How is it that some mammals are nocturnal?"

These capacities include highly developed olfactory abilities (both high sensitivity and acute powers of discrimination) often coupled with an elaborate system of scent marking which has been shown capable of mediating, in darkness, important aspects of the behaviour of terrestrial and underground mammals [see, for example, Muller-Schwarze and Mozell (1977), Albone (1984) and Epple (1986) for summaries and review]. In many mammals hearing often extends to sound frequencies quite beyond our own audible range (ultrasound) and this, coupled with an ability to produce high frequency sounds, may be used for communication between individuals (Sales and Pye 1974) and more importantly to detect the presence of objects by the mechanisms of echolocaton (Griffin 1958; Purves and Pilleri 1983). The high frequencies of these sounds are essential for the detection of fine detail and this ability has been shown to be highly developed in such diverse mammalian orders as the dolphins (Cetacea) and bats (Chiroptera), enabling species from both groups of animals to orient their movements in three dimensions by this means alone.

In many mammal species (for example among the Rodentia and the Insectivora) both the ability to detect ultrasound and high olfactory sensitivity occur together. Just how these senses work together to cope with the sensory problems faced by the nocturnal mammals is not understood in a detailed way, but at least possession of these sensory capacities would seem to form the basis for explaining in general terms the differences between many mammals and ourselves as regards mobility under the cover of darkness.

Can the behaviour of the nocturnal birds be explained in a similar way? Do they have any sensory capacities additional to those of diurnal birds, or even additional to those of ourselves, which can account for their nocturnal activity? If the answer to these questions is negative, how can the nocturnal activities of birds be explained? Almost any general text in ornithology will point out more or less as an "obvious fact" that the behaviour of birds is guided primarily by vision. Rochon-Duvigneaud (1943) encapsulated this idea in the phrase, "A bird is a wing guided by an eye". But can vision account for the behaviour of nocturnal birds? What of the role of hearing, olfaction or touch sensitivity? As will be discussed below, vision is not the only way to guide a wing. The echolocatory skills of the bats attest well to the fact that a wing can in fact be guided by an ear. However, only a bare handful of birds have the ability to guide their flight by echolocation. The majority of nocturnal birds cannot echolocate. But, can these nocturnal birds still be characterised as a wing guided by an eye?

The main purpose of this book is to examine these questions in detail and this will be done primarily by reference to one particular group of birds in which complete nocturnality is relatively common, the owls (Strigiformes). However, before doing so it is essential to provide a context in which to view the activities of these strictly nocturnal birds.

First will be considered some activities of those species of birds which are not strictly nocturnal but are nevertheless sometimes (often regularly) active outside of daylight hours, including dusk and dawn. Second, the rather special cases of the birds which fly, nest and roost within the completely dark interior of caves will provide a useful standpoint from which to view examples from the third group of birds, the truly nocturnal species. Discussion of these will begin with the flightless nocturnal birds and conclude by considering instances of true nocturnality among flying birds.

In order to provide a full context in which to consider the sensory and behavioural problems faced by the nocturnal birds, the actual light levels associated with the daily cycle, through daylight, twilight and into night-time, must be discussed. It will be seen that the light levels of night-time and twilight are far from uniform and depend upon both geophysical factors such as latitude, season, the lunar cycle and weather, and upon biological factors associated with different habitat types and the seasonal changes which they undergo. The final link will be to consider what is known of the sensory capacities of the owls and other birds, how they are limited by physical and physiological factors and how these sensory capacities match up to the problems posed by the natural nocturnal environment. Only then will it be possible to consider the question, "How is it that owls are nocturnal?"

CHAPTER 2

What is night-time?

EXPERIENCING NATURAL NIGHT-TIME

As remarked in the introduction, man is, or usually behaves like, a diurnal animal. This means that most of us rarely experience "natural night-time" with the result that we are unaware of what the natural night-time environment actually consists of. Indeed it may be very difficult to appreciate true night-time conditions in our daily lives, for if we live in or near a large town or city, artificial sources of light obscure true night-time. Even in rural areas the influence of lights from quite distant towns and cities may still be detected, especially during cloudy weather when the light from a town may be reflected off the base of clouds. It is especially under these conditions that the distant glow of a city may be seen many miles away. Such "light pollution" of the night sky from major cities can render the sky many times brighter than the natural minimum background and it is a major source of concern to astronomers. For their purposes the light pollution from cities up to 100 kilometres away can significantly raise the background light levels of clear night-time

7

skies and there are international pressure groups who campaign for "natural night skies" (Henbest 1989).

It is to be hoped that any readers not familiar with the night environment will be tempted to experience for themselves something of the natural night environment, though as explained below the experience of a single night, or even many nights at the same latitude, will not be sufficient to appreciate fully all that night-time can present.

Even if a suitable site well away from the effects of artificial lights can be found whereby true night-time can be experienced two further problems arise. First, it is inappropriate to try to experience night-time instantaneously; time is needed to allow vision to "dark adapt" to the ambient light level, and secondly, what we colloquially refer to as night-time is far from uniform; the light levels experienced "at night" will vary considerably with time of year, latitude, and weather conditions such that an "average night" becomes impossible to experience or describe. However, it is important to come to some understanding of these problems because it is against this background of varying night-time light levels that the nocturnal activities of animals have evolved.

ADAPTING TO THE DARK

To appreciate what we may be capable of detecting visually at night requires us to be exposed continuously to night-time light levels for an extended period. The minimum time is about 40 minutes, for it is only after such a length of time that our vision will have become fully "dark adapted", that is, the visual system will have achieved its maximum sensitivity.

That our eyes slowly increase their sensitivity with continued exposure to darkness has long been understood and it is something that we can easily verify for ourselves. The first systematic investigations of dark adaptation were conducted over a century ago (Aubert 1865). Not only was it demonstrated that visual sensitivity increases with continuous exposure to the dark but it was also shown that even a brief exposure to a bright light source during the process of dark adaptation will reset the eye's sensitivity to a lower level. Following even a brief exposure to a bright light the process of dark adaptation must begin over again if high visual sensitivity is to be attained. Thus there is no quick way of appreciating our ability to detect things visually at night; it is essential to wait for the processes of dark adaptation to take their full course. It is not sufficient to drive to a remote spot, switch off the vehicle lights and make a quick assessment of what can be seen at night. At least 40 minutes exposure to the ambient conditions would be necessary to gain some insight into the problems that the surrounding night actually presents.

In addition, dark adaptation seems to be a property of all vertebrate eyes. The first investigation of dark adaptation in a bird, the Pigeon or Rock Dove *Columba livia*, indicated that sensitivity increases as a function of time in the dark in a way that is very similar to that found in humans (Blough 1956). Thus, it is necessary to be very cautious in making comparisons of visual capacities at night between ourselves and other animals based upon casual field observations.

Even if the visual tasks which are being compared are directly equivalent it is highly unlikely that the state of dark adaptation will be the same in both species. Indeed while 40 minutes is a good rule of thumb for the full time course of dark adaptation, the speed with which sensitivity increases and its final level are both dependent on the brightness and duration of the light that an observer is initially exposed to. For example, a day spent in bright sunlight on an exposed beach is sufficient to retard the full extent of dark adaptation for a period of several hours, or even days (Hecht *et al* 1948). After 20 minutes in the dark two observers may differ in their visual sensitivity by over ten-fold, depending on the brightness of the light they were initially exposed to. [See Bartlett (1965) and Barlow (1972) for detailed reviews of how the sensitivity of the eye during dark adaptation changes due to differences in the preadapting light conditions.]

The importance of this point can be appreciated when it is realised that the full range of dark adaptation increases the overall sensitivity of the eye by more than 10,000-fold (4 log units). Following exposure to low day-time light levels, sensitivity increases by an average of about 100-fold in the first 10 minutes of exposure to darkness, and by a further 100-fold over the next 30 minutes, though the rate of increase of sensitivity in these two periods is not uniform. Clearly such marked changes in visual sensitivity over time are not trivial and it can be readily seen therefore that any comparisons between species in terms of their visual sensitivity (or ability to distinguish between brighnesses) must involve carefully controlled periods of dark adaptation.

It has been argued (Lythgoe 1979) that the rate of dark adaptation is sufficient to keep pace approximately with the most rapid natural changes in light levels which occur as day changes into night through the twilight period. Thus by exposing oneself to the natural changes of light levels at the end of the day, vision is likely to be always at its most sensitive for the ambient light levels which prevail through twilight. Indeed it seems likely that in the natural world it is only by plunging oneself under dense vegetation or into the darkness of a cave that the sensitivity of vision will not be well adapted to the ambient light levels.

It is important to be well adapted to the ambient light level and not to have one's vision caught in the process of dark adaptation as humans often are, following the extinguishing of a bright light source, such as a torch. This is indicated by the fact that contrast discrimination (the ability to distinguish between the brightness of two lights presented simultaneously) is at its most acute when the eye is fully adapted to the background intensity (Barlow 1962). In other words we are best able to appreciate differences in the brightness of various objects in the scene if the eye is adapted to the overall ambient light level. [See Brown and Mueller (1965) for a discussion of some of the basic parameters which influence our ability to discriminate between light sources of different brightness.]

Since it is this discrimination between brightnesses which is the cue to the detection of objects, an animal should ideally attempt to remain well adapted to the ambient light and eschew exposing itself to sources of light which are appreciably brighter than the ambient. Human beings in the modern world are perhaps the only animals which voluntarily seem not to do this, readily switching from brightly lit situations to darkness. However, we can easily appreciate the problem when gazing from the camp fire into the surrounding night-time gloom. The effect of gloom is more a product of our inability to detect brightness differences between the objects around us because of an inadequately adapted eye, rather than a property of the night environment itself. If we stare fixedly at the gloom it will slowly fade as the eyes become dark adapted and objects begin to appear as our ability to detect brightness differences is enhanced.

However, it must be noted that although objects "appear out of the gloom" as the eye becomes adapted to the ambient light levels, the amount of detail that can be discerned will be a function of the ambient light level. Fine detail will not be evident at low light levels even though the eye has become well adapted to that light level. The amount of detail which can be discerned in a scene, the eye's visual acuity, is also a function of light level. This is discussed in more detail later, but it is important to note here that optimal acuity at a given light level is only achieved if the eye is well adapted to that light level.

THE VARIABILITY OF NIGHT-TIME LIGHT LEVELS

It is possible both to choose a site well away from artificial light sources, and to allow for the processes of dark adaptation, and thereby experience something of the natural light regime of night-time. However, there is one further problem which makes it difficult to appreciate easily what the night environment consists of, or even to define "night-time". This is the fact that the light levels of night-time are far from uniform.

The light regime experienced at the earth's surface throughout any 24 hours is a function of many variables which can cause light levels to vary over a phenomenally wide range. The light regime experienced on any one night will provide just a single sample of what night-time can consist of. More importantly the natural light levels of night-time are more variable than those of daytime. However, it must be appreciated that it is these variable light levels which nocturnally active animals will have evolved to cope with rather than some arbitrarily defined level which may be labelled "darkness" or "night-time".

In temperate and tropical latitudes light levels experienced in open habitats may vary during a 24-hour period over a range of approximately three billion-fold (9.5 log units). If account is taken of the shading that a tree canopy provides then the total range of light levels which can be experienced at the earth's surface within 24 hours (without going into the complete darkness of a cave) will be approximately 100 times greater, i.e. 300 billion-fold (11.6 log). These figures are so large it is difficult to appreciate what they mean and Figure 2.1 presents some of this information in graphic form. However, because the differences in light levels are so large the data must be presented on logarithmic scales where a difference of one unit represents a change in light levels of ten-fold, or one order of magnitude.

Figure 2.1 and Table 2.1 show the average illumination levels (in lux) which are produced at the earth's surface by the full range of natural light sources;

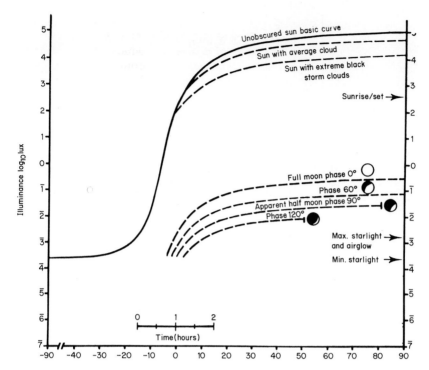

Figure 2.1 The total range of natural illumination levels (log lux) in open habitats. The illumination levels due to the sun, as a function of altitude, is indicated by the "unobscured sun basic curve". The range of illumination levels due to the moon at different phases and as a function of altitude are indicated, together with the maximum and minimum levels which can be produced by starlight alone. The illumination level produced by the sun at sunrise or sunset is also indicated. The basic curve is for latitude 50° at the time of the summer solstice. At lower latitudes the transition from night- to day-time is more rapid and is fastest at the equator, but the absolute differences in light levels are similar to those shown here. At higher latitudes the transition from day- to night-time is slower and the total range of illuminance experienced on both a daily and annual basis is less than at lower latitudes (days are less bright and nights are less dark at higher latitudes). At the poles there is no variation of illumination with each day, light levels change only slowly from day-to-day. The data are based upon the Natural Illumination Charts of the US Navy (1952).

the sun, moon, stars and airglow from the atmosphere. They are based upon the Natural Illumination Charts of the US Navy (1952) and on the data of Bond and Henderson (1963). The light levels produced by the sun are a function of its altitude with respect to the horizon. It can be seen that light levels change most markedly in the period when the sun travels between 10° above and 10° below the horizon. For most of the day-time, however, the light levels produced directly by sunlight change by only approximately ten-fold.

TABLE 2.1. Summary of natural illumination levels, based upon the US Navy Natural Illumination Charts (1952).

Condition	Limit	Sun's altitude (degrees)	Illumination	
			lux	log lux
Daylight	upper	+90	123786	5.09
	lower	−0.8	452	2.66
Twilight: Civil	upper	−0.8	452	2.66
	lower	−6	3.4	$\bar{0}.53$
Nautical	lower	−12	0.00829	$\bar{3}.92$
Astronomical	lower	−18	0.000646	$\bar{4}.81$
"First" or "last" light:		−8	0.112	$\bar{1}.05$
Moonlight	upper	(full moon at 90 degrees)	0.371	$\bar{1}.57$
	lower	(quarter moon at 22 degrees)	0.0133	$\bar{2}.12$
Starlight	upper		0.0108	$\bar{2}.03$
	lower		0.00030	$\bar{4}.48$

This is despite the fact that the sun's altitude in the sky may change dramatically.

Cloud cover will, of course, alter ambient light levels but even though photographers may be driven to despair by the light level changes produced by cloud these changes are almost trivially small compared to the overall changes in ambient levels which occur during the twilight periods. Thus, the thickest storm clouds only reduce light levels by about ten-fold and "average" cloud cover reduces light levels by a factor of only two- or three-fold.

Thus in an open habitat, throughout most of the day-time, light levels are relatively uniform, indeed during most of the day more variation in light levels can be produced by a vegetation canopy than by variation in the sun's altitude and average cloud cover combined [e.g. Federer and Tanner (1966) estimate the mean attentuation of light produced by a continuous broad-leaved tree canopy to be approximately 100-fold].

Light level changes during night-time, however, are more complicated than during the day. This is mainly because of the complex way in which the moon changes both in its phase as well as in its altitude. The light levels produced by moonlight may vary over a range of approximately 1.45 log units (28-fold). Even different sections of the moon's disc vary in their reflectance so that the moon at its first quarter (waxing) is about 20% brighter than at the third (waning) quarter. Also the distance of the moon from the earth varies during the lunar cycle such that light levels due to this influence alone will affect light levels by about 26%.

The moon is often not present at night and in its absence ambient light is provided by the stars and airglow. The ambient light produced by the stars themselves varies and is a function of both geographical position and time of year. Airglow is present both night and day but its contribution to the ambient light levels is so small that it only becomes significant on moonless nights. Airglow, which is not the same as the aurora experienced mainly at very high latitudes, is attributable in the main to reactions between free atoms of oxygen between 70 and 100 km above the earth's surface. Some of these reactions may also involve charged atomic particles from outer space. The actual light levels produced may change from night-to-night and with season. These variations in airglow, coupled with the variations in starlight can cause ambient light levels to vary over a range of approximately 1.55 log units (35-fold).

The attenuation produced by cloud cover and vegetation will be the same for moonlight and starlight as they are for sunlight. Figure 2.1 shows that although moonlight and starlight are responsible for light levels which vary over ranges of approximately equal size, the two ranges do not overlap. Thus, the light levels produced by the two principal night-time light sources, the moon and the stars, can cause natural illumination levels to vary over a range of up to 1,200-fold from night-to-night, or even within one night should there have been full moonlight followed by minimum starlight once the moon had set. Variable cloud cover within the night could increase this variation to 12,000-fold.

DEFINING NIGHT-TIME LIGHT LEVELS

When does night-time begin in terms of light levels and the position of the sun? As is made clear from our own experience of dusk and dawn, and from the data of Figure 2.1, the transition between night and day is characterised by continuous and relatively slow changes in light levels; there is no abrupt transition in nature between night and day. Therefore, arbitrarily defined limits based upon the position of the sun with respect to the horizon are now used to define the various stages of the transition between night and day. In the natural world these arbitrarily defined limits are probably of little functional significance but they are in common usage and can act as useful guides for describing how the light regimes which constitute night-time change with latitude and season. The following definitions are now internationally agreed. The actual illumination levels associated with the limits of these phases of night and day are given in Table 2.1.

Daylight: This extends from the time of sunrise to the time of sunset. Sunrise/set are the instants when the top of the upper limb of the sun's disc is seen on the horizon. At this moment the sun's disc is actually completely below the horizon but because of the refraction of light through the atmosphere the sun appears higher than it actually is. At sunrise/set the centre of the sun's disc is 0.8° below the horizon.

Twilight: These are the light intensities which occur when the sun is within certain ranges below the horizon. Light at this time is caused by the scattering of sunlight from the upper layers of the atmosphere; that is, twilight strictly refers to indirect illumination from the sun. There are three twilight zones, the upper limit of all three zones is sunrise/set, the lower limits are (1) *civil twilight*, centre of sun's disc 6° below the horizon; (2) *nautical twilight*, centre of sun's disc 12° below the horizon; (3) *astronomical twilight*, centre of sun's disc 18° below the horizon. When the sun's position is at or below the lower limit of astronomical twilight, the contribution from the sun's scattered light to the total ambient illumination is less than the contribution from starlight, and of the same order as that from airglow. Thus, it is not until the end of astronomical twilight that it is considered that the sun no longer contributes to night-time light levels.

Moonlight: This is arbitrarily considered to be the condition between when a quarter moon is approximately 22° above the horizon to when the full moon is at its zenith. It can be seen from Table 2.1 that this moonlight range actually falls within the lower half of the range of light levels produced during nautical twilight.

Starlight and airglow: This is simply the range of naturally occurring light levels below minimum moonlight and is in fact due to a number of sources other than just the light from stars. These include airglow and zodiacal light. The latter is mainly attributable to the light given off by meteors as they burn up in the atmosphere. The number of meteors entering the atmosphere is not trivial (an estimate of 10^{10} per annum) and these do provide some contribution to ambient light at these low levels.

It should be noted from Table 2.1 that there is a definite minimum which marks the lower limit of natural illumination of the earth's surface in open habitats. By entering beneath a vegetation canopy this level can be reduced by more than 100-fold. However, only by deliberately attempting to reduce light levels still further, by entering a cave, for example, can complete darkness be appreciated. In an open situation complete darkness can never occur.

As mentioned above, and also made clear in Figure 2.1, the transition between night and day is a smooth one and it is clearly not possible to decide other than by some arbitrary criterion that day is over and night-time has begun. It is especially important to bear this in mind when considering the activity of animals in relation to night-time. For example, two different field observers may describe the activity of an animal as "beginning at dusk". However, because light levels can be so variable and change so rapidly through the twilight period, it is possible that the observers will in fact be referring to quite different light levels, perhaps differing by up to 1,000-fold or more. Clearly this does raise important problems when analysing the nocturnal behaviour of animals or even when describing the factors which may be associated with a diurnally active animal ceasing its activity at the end of the day.

There is no simple solution to this problem other than to expect observers to measure the ambient light levels at the time observations are made. Clearly this is not feasible for most field observers and has in fact been done in only a handful of studies. The most important point to note is that even during the middle of the night (well outside the various twilight periods), "night-time" can embrace at least a 1,000-fold range of light levels (maximum moonlight to minimum starlight, under clear skies), and that the transition through the twilight period at the beginning or the end of each night will always embrace a further 1,000-fold range of light levels (sun rise/set to maximum moonlight). If account is taken of the difference in light levels which are produced by moving between an open habitat to under a tree canopy then light levels in each of these above ranges may vary by a further 100-fold.

Clearly, defining "night-time" is difficult. There is no simple answer or value for a night-time or twilight light level. It is necessary to think in terms of a *range* of light levels in which an animal may be active. Although this may seem difficult or unsatisfactory it should be borne in mind that it is with such a range of light levels that the nocturnally active animal has to cope on a night-to-night basis. Its behavioural and sensory adaptations will have evolved within the context of these widely varying night-time light levels, and so it is within this context that the behavioural and sensory adaptations which allow nocturnal mobility are to be sought.

NIGHT-TIME, LATITUDE AND SEASON

There is one further complicating factor in providing a full description of night-time and the transition to it from day-time through dusk. This is the effect of latitude. It has major influences on the length of both night-time and the twilight period. It also influences the maximum and minimum light levels which are experienced at a particular site at a particular time of year.

For example, at some latitudes at certain times of the year night-time never occurs; at other latitudes light levels do not fall below those of twilight for many weeks, while at other latitudes the length of night-time and twilight hardly change throughout the year. This means that on the same calendar day a nocturnally active animal at one latitude may experience a quite different light regime to that experienced by a nocturnally active animal at another latitude; alternatively an animal nocturnally active at a particular latitude may experience virtually the same light regime every night throughout the whole year while an animal at a different latitude will experience marked changes with season in the duration of the night and twilight.

Strictly, the data of Figure 2.1 applies to a latitude of 50° (the latitude of the southernmost part of the British Isles, southern Argentina or the city of Vancouver, Canada) at about the time of the summer solstice (June 21st). The illumination levels indicated will vary little when moving towards the equator. However, moving towards the poles influences both night-time light levels and the speed of transition through dusk in a dramatic way. The difference in the maximum light levels ever experienced at mid-day between the equator and the poles is approximately three-fold. That is, to within a small degree, it is as bright at mid-day all over the world. But how dark it is at midnight varies very markedly with both latitude and season.

Probably the best way to appreciate the effect of latitude on naturally occurring light levels is to describe the situation at the equator and then consider how this is altered by moving towards the poles. Figures 2.2, 2.3 and 2.4 illustrate graphically how night-length, and the length of civil and nautical

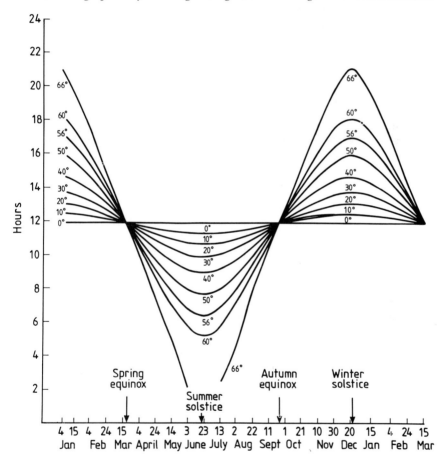

Figure 2.2   The variation of night length (defined as sunset to sunrise) as a function of time of year, at different latitudes. Each curve indicates how night length changes at a particular latitude over a 14-month period. The latitude for each curve is indicated, as are the times of the solstices and equinoxes. Around the time of the summer solstice the sun does not set at latitudes above 66° and there is no night-time. Based upon data in *The Astronomical Almanac* (1988).

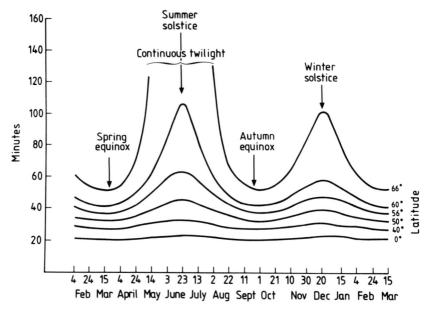

Figure 2.3   The length of civil twilight (the period when the sun is between 0°
and 6° below the horizon) as a function of the time of year at different latitudes.
Each curve indicates how the length of civil twilight changes at a particular
latitude over a 14-month period. Around the time of the summer solstice, at
latitudes above 66° twilight does not turn to night. From May 14th–June 11th,
and from July 1st–29th, light levels never fall below that of civil twilight, and
from June 15th–27th the sun does not set. Based upon data in *The Astronomical
Almanac* (1988).

twilight change with time of year at different latitudes. They can be used to
find out how these variables change at a given latitude throughout the year or
to see how these variables differ on the same day at different latitudes. The
data is based upon *The Astronomical Almanac* for 1988. This is an annual
publication which should be consulted for more specific details of night-
length, etc, at different latitudes throughout each year.

At the equator (latitude 0°) both night-length (Figure 2.2) and the speed of
transition between night and day, indicated by the length of civil and nautical
twilight (Figures 2.3 and 2.4), alter little with season. Night-time (between
sunset and sunrise) is within a minute of 11 hours 53 minutes long throughout
the year; the transition between night and day (indicated by the length of civil
twilight) is always within 90 seconds of 21 minutes and is the shortest civil
twilight period experienced anywhere in the world. Sunset and sunrise occur
at the same time every day.

Thus a nocturnally active animal living at or near the equator experiences,
for all intents and purposes, an identical light regime (in terms of night and

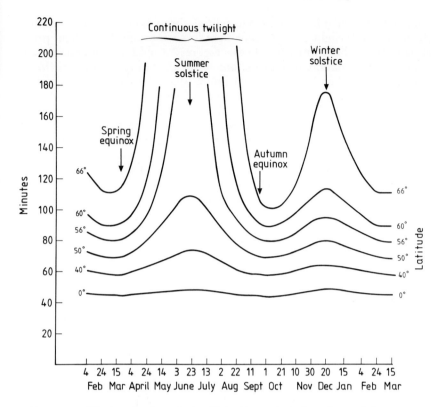

Figure 2.4 The length of nautical twilight (the period when the sun is between 0° and 12° below the horizon) as a function of time of year at different latitudes. Around the time of the summer solstice, light levels are continuously above those of the lower limit of nautical twilight for varying periods, depending upon latitude. At latitude 56° this period lasts from June 3–July 9th, at latitude 60° from May 14th–July 29th, and at latitude 66° from April 24th–August 18. Based upon data in *The Astronomical Almanac* (1988).

twilight length) every day of the year. There, night-time is predictably the same from one night to the next.

As an observer moves away from the equator towards the poles three major changes occur: (1) night-length changes markedly with the season (Figure 2.2); (2) the speed of transition between night and day (i.e. twilight-length) changes markedly with the season (Figures 2.3 and 2.4); and, (3) the illumination experienced at midnight (independently of the moon) on each day is dependent upon season.

Consider first the situation at latitude 40° (approximately that of the island of Mallorca in the Mediterranean, Washington DC, Peking, or Wellington, New Zealand). Here the overall range of illumination throughout each 24 hours is very similar from one day to another throughout the year. The rate of change from daylight to night-time light levels through the twilight period also varies very little with time of year (Figures 2.3 and 2.4). For example, civil twilight lasts a minimum of 24 minutes at the time of the equinoxes (March 21st and September 23rd) and a maximum of 33 minutes at the time of the summer solstices (June 21st in the northern hemisphere, December 22nd in the southern hemisphere). What does alter quite significantly at this latitude is the time from day-to-day of sunrise and sunset such that night-length is progressively changing. It can be seen from Figure 2.2 that night-length (sunset to sunrise) is only 9 hours at the summer solstice and rises to 14 hours 40 minutes at the winter solstice. If night-time is defined as extending only between the lower ends of nautical twilight then night-length at this latitude varies between 6 hours 30 minutes and 12 hours 30 minutes. The importance of such changes in night-length is that at the same time of day depending upon season, the light level experienced at any one place on the 40° latitude can vary by many million-fold, i.e. the same time can be either full daylight or starlight depending upon season.

These changes in the length of the night and of the twilight periods with season, which are already noticeable at latitude 40°, become increasingly large further from the equator. Figures 2.2, 2.3 and 2.4 show that these differences are much greater between latitudes 50° and 60° than they are between 40° and 50°.

As far as night-time is concerned one of the most significant changes is that at high latitudes night-time may never occur around the time of the summer solstice. Figure 2.4 shows that at this time of year light levels never fall below those of nautical twilight at a latitude as low as 56° (Glasgow, Labrador, Cape Horn) for a period of 36 days. It can also be seen that light levels below those of civil twilight do not occur for many weeks at latitudes above approximately 60°. [At latitude 66° light levels do not fall below those of civil twilight between May 14th and July 29th (11 weeks), furthermore, there is continuous sunlight for 12 days even at this latitude.]

These observations are quite pertinent to the theme of this book since many bird species, both passerine and non-passerine, breed and migrate at these latitudes in the northern hemisphere during the summer months. For example, in the Western Palearctic over 200 bird species (131 non-passerine, 72 passerine) breed regularly at the latitude of the arctic circle (66.5°) or higher (Harrison 1982), and in the Nearctic 219 species (177 non-passerine, 42 passerine) breed at these same latitudes (National Geographic Society 1983). While the majority of these species are migrant summer visitors, a not insubstantial proportion also stay at this latitude throughout the winter as well. Of the Western Palearctic species, 62 may be resident throughout the year at this latitude, and 26 of the Nearctic species may do so.

It is worth nothing that nautical twilight alters by a larger percentage than civil twilight over the annual cycle at all latitudes. Also, while at the equinoxes night-length (sunset to sunrise) is the same at all latitudes (11 hours 53 minutes, Figure 2.2) and the twilight periods at this time are the shortest of the year (Figures 2.3 and 2.4), they are nevertheless of very different durations depending upon latitude. Thus Figure 2.3 shows that at the equator civil twilight at the equinoxes lasts just 22 minutes, at latitude 50° it is 32 minutes long and at latitude 66° it is 51 minutes, despite the fact that night-length is the same at all latitudes on these days.

The most dramatic changes in light regime with latitude are apparent at locations above the arctic or antarctic circles. At a latitude of approximately 76°N [the Svalbard Archipelago, Thule (Greenland), The Queen Elizabeth Islands (Canada)], which is 10° beyond the arctic circle, over 30 Western Palearctic and 40 Nearctic bird species regularly breed. These are mainly migratory species of sea birds (principally gulls, terns and auks, shorebirds and waterfowl), but they also include land birds such as the Snow Bunting (*Plectrophenax nivalis*), Gyrfalcon (*Falco rusticolus*), Snowy Owl (*Nyctea scandiaca*), Ptarmigan (*Lagopus mutus*), Red (or Willow) Grouse (*L. lagopus*) and Raven (*Corvus corax*). Individuals of the latter five species may sometimes be resident throughout the year even at this latitude [see comments on the Raven in Goodwin (1986), on the Snowy Owl in Cramp (1985), and on the Gyrfalcon, Ptarmigan and Red Grouse in Cramp and Simmons (1980)].

At a latitude of 76° the annual light regime has the following features. Light levels change by only a small amount over very long time periods, especially in midsummer and in midwinter, around the solstices. In summer there is seemingly endless light while winter is characterised by apparently endless darkness. The sun stays continuously above the horizon for almost a third of a year from April 24th to August 19th (118 days). At such high latitudes the sun appears to go around the horizon in a gentle sloping plane rather than steeply up, over and down as is apparent at lower latitudes. In midsummer the light from one day runs into that of the next without any appreciable change. In fact the difference in light levels between "midnight" and "midday" may only be about six-fold. At the other extreme is the 6 week period from December 1st to January 13th when the sun never climbs above an elevation of 8° below the horizon and so light levels never get above those of the moonlight range.

At this latitude (76°) "days" throughout the year can be characterised as falling into four broad categories:

1. Daylight only, with the sun continuously above the horizon, from April 24th to August 19th (118 days).
2. Sunlight and twilight only; this occurs at two periods from August 20th to October 5th and again from March 10th to April 23rd (92 days).
3. Sunlight, twilight and night; also for two periods, October 6th to 31st and February 11th to March 9th (54 days).
4. Twilight and night only between November 1st and February 10th (102 days).

For completeness it is worth describing briefly the annual pattern of illumination at the poles, since this presents the other extreme to the very regular pattern of natural illumination changes experienced every 24 hours at the equator. At the pole on any given day the light changes not at all, save for the influence of the moon or cloud cover. At all times the sun, when it is visible, appears to stay at one place in the sky. The sun does appear to go around the horizon but never up, over and down. It takes nearly four days for the sun to rise and to set at the spring and autumn equinoxes. At the north pole the annual cycle is approximately as follows:

Sunrise (spring equinox):	March 20th–March 23rd	4 days
Continuous sunlight:	March 23rd–September 20th	182 days
Sunset (autumn equinox):	September 21st–September 24th	4 days
Civil twilight:	September 25th–October 9th	15 days
Nautical twilight:	October 10th–October 25th	16 days
Astronomical twilight:	October 26th–November 13th	19 days
Starlight and airglow:	November 14th–January 29th	77 days
Astronomical twilight:	January 30th–February 17th	18 days
Nautical twilight:	February 18th–March 5th	16 days
Civil twilight:	March 6th–March 19th	14 days

These examples of the different light regimes which may be experienced throughout each 24 hours at four different latitudes serve to illustrate some important points about what night-time actually consists of in different parts of the world at different times of the year. As has been seen, away from the equator the time between sunset and sunrise is far from uniform, as also is the length of time that can be called night. It is worth stating three main points which derive from the above discussion and description of night-time.

First, it is incorrect to attempt to refer to the quantity of light in broad terms by reference to time of day only. For examle, 30 minutes before dawn the light levels experienced can differ over a range of 500-fold depending upon the time of year and the latitude.

Secondly, days cannot be divided simply between day-time and night-time. Time of year, latitude, as well as the hour of day, produce wide variations in the altitude of the sun and hence the light levels experienced. There is a wide variety of types of day and types of night as far as light is concerned. From the equator to the poles, days can be all dark, some all twilight while others show huge variations in light regime.

Thirdly, it is wrong to consider one value for starlight, one for moonlight, one for twilight, etc. In each case the illumination under these headings ranges over a wide scale.

LUMINANCE AND ILLUMINATION

One final point that needs discussing briefly, here, is the distinction between the illumination falling on a surface and the brightness or luminance of that surface. The actual definition and measurement of light is a complex task (see Wyszecki and Stiles (1967), Walsh (1958) or Riggs (1965)) and it has been necessary to devise two parallel systems of measurement: *radiometric units* which measure light purely in terms of energy, and *photometric units* which measure light in terms of its effectiveness as a stimulus for vision. In the final analysis, however, photometric terms and units may be related mathematically to radiometric terms and units. In all of the discussion in this book photometric units will be used and we need to distinguish only between *luminance* and *illuminance*.

Illuminance measures, as presented in Figure 2.1 and Table 2.1, quantify the amount of light which falls onto, or illuminates, a given surface area. *Luminance* measures, give an indication of the brightness of a surface when it is observed. The two measures are related under defined circumstances such that an illuminance of 1 lux (or 1 lumen/m^2) falling on an "ideal" surface will result in that surface having a luminance of $1/\pi$ nit (or $1/\pi$ cd/m^2). (An ideal surface is one that is white, perfectly reflecting and perfectly diffusing; a sheet of ordinary, white cartridge paper comes close to this ideal but it is most closely approximated by the white deposit left on a smooth metal surface when held in the smoke of piece of burning magnesium ribbon.)

The important difference between luminance and illuminance is that when referring to the illuminance of a surface, that surface is then being considered as a *receiver* of light energy; when referring to the luminance of the surface it is being considered as a *source* of light energy which acts as the stimulus for vision. Illuminance will therefore change as the surface is moved nearer or further from the source of light illuminating it; on the other hand, a surface having a given luminance will be unchanged regardless of the distance from which it is viewed.

Two different surfaces receiving the same illumination may, however, reflect the incident light which falls on them with different degrees of efficiency. This means that two different surfaces, each receiving the same illumination, will have different luminances. Indeed it is these differences in the reflectance of surfaces which leads to differences in the visual contrast of objects illuminated by the same light sources whether that source is the sun, moon or an electric lamp. It is for this reason that luminance units are preferred when describing the visual problems which face an animal in the natural world. They can be compared directly with measures of the visual capacities of an animal.

All objects in a given scene are illuminated by the same natural light sources (sun, moon, stars, sky) which are so far away that effectively all objects in a scene receive the same illumination. However, these objects differ in their reflectance and hence luminance.

Table 2.2 and Figure 2.5 present the average luminance of two different surfaces (an "ideal" surface and grass or leaf litter) which result when illuminated by the various different natural light sources as described in Table 2.1 and Figure 2.1. It thus presents the luminance that we or a bird would perceive when looking directly down onto the surface below us. Since these are luminance measures it makes no difference whether the surface is observed from just a few metres above the ground, from the top of a tree or from a thousand metres as a bird flies over.

In constructing the figure the natural substrate is assumed to have a reflectance of 25%, which is about equal to that of grass or leaf litter. However, different natural objects in the scene will have different degrees of reflectance. Therefore while Figure 2.5 indicates the average luminance of the ground, specific objects will be revealed by how their degree of reflectance

TABLE 2.2. Summary of luminance levels produced under different conditions of natural illumination on an "ideal" surface (white, perfectly reflecting and perfectly diffusing) and on a natural substrate, such as grass or leaf litter, which has a reflectance of 25%. Based upon the natural illuminance data of Table 2.1.

Condition	Limit	Sun's altitude (degrees)	Luminance (log cd/m²)	
			ideal	grass
Daylight	upper	+90	4.59	3.99
	lower	−0.8	2.16	1.56
Twilight: Civil	upper	−0.8	2.16	1.56
	lower	−6	0.03	$\overline{1}$.43
Nautical	lower	−12	$\overline{3}$.42	$\overline{4}$.82
Astronomical	lower	−18	$\overline{4}$.31	$\overline{5}$.71
"First" or "last" light:	lower	−8	$\overline{2}$.55	$\overline{3}$.95
Moonlight	upper	(full moon at 90 degrees)	1.07	$\overline{2}$.47
	lower	(quarter moon at 22 degrees)	$\overline{3}$.62	$\overline{3}$.02
Starlight	upper		$\overline{3}$.53	$\overline{4}$.93
	lower		$\overline{5}$.98	$\overline{5}$.38

differs from that of the background. In nature, however, the degree of reflectance of many natural objects and surfaces (e.g. leaf litter, tree bark, bare soil, fresh and dried grass) do not in fact differ very markedly (Wyszecki and Stiles 1967). Indeed the essence of natural camouflage is the matching of an animal's reflectance to that of the substrate against which it is most likely to be viewed by a predator. Snow covered countryside will clearly have a higher luminance than grass or bare soil under the same illuminating conditions due to its higher reflectance and may in fact come close to the "ideal" values of Table 2.2. However, many natural surfaces have a reflectance similar to that of grass or leaf litter which was used in computing the luminances of "average ground" as given in Figure 2.5.

Natural substrates are also highly diffusing surfaces. That is, they are surfaces from which incident light is reflected or scattered in many different directions simultaneously. The effect of this is that the luminance of the surface will change little over a wide range of angles of view about the perpendicular to the surface. In contrast with this are smooth or polished surfaces, such as water, which may be highly reflective but not diffusing. Here, the angle from which the surface is viewed relative to the source of illumination can considerably alter its brightness. This phenomenon is, of course, the source of highlights seen on any smooth surface, the actual position of the highlight changes with the angle of view.

Figure 2.5 also shows the luminance of the sky near the horizon under different natural conditions. This gives some indication of how bright the sky will appear to an observer. The luminance of the sky can be compared directly with the luminance of the ground under the same ambient conditions. It can be seen that in open habitats, the luminance of the sky near the horizon is often very close to that of the average ground, or grass, when viewed from above, under the same conditions of natural illumination. A snow covered landscape could be brighter than the sky near the horizon under the same conditions.

Figure 2.5 shows the average luminance of this same surface under four different conditions. (A) in the open without cloud cover, (B) in the open but with maximum cloud cover, (C) under a continuous woodland canopy but without cloud and (D) under a trees canopy with maximum cloud cover. Moving beneath a vegetation canopy can reduce the luminance of the ground by up to 100-fold and in this situation the sky, if visible through the canopy, will then be brighter than the substrate. Thus it can be seen that there can be up to 1,000-fold difference in the light level experienced at (say) full moonlight depending upon the presence of cloud and whether the observer is beneath a woodland canopy.

Outside of the canopy objects may be made more perceptible by viewing them against the sky as a silhouette. This is because when viewed from below or perpendicular to the direction of the incident illumination, the luminance of an object will be considerably reduced, compared with viewing it from above or from a moderate angle to the illumination. Thus viewed, the luminance of the object may be noticeably below that of the sky towards the horizon and hence an object will become visible as a dark object against a brighter background.

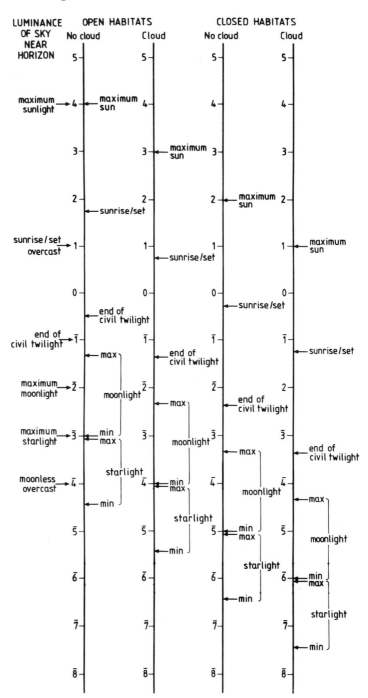

Figure 2.5 indicates therefore the likely range of natural luminances of both the ground and the sky, which an animal could experience depending upon whether it is inside or outside a woodland and upon what the weather conditions are. Viewing dark soil, such as wet peat, or viewing a surface of freshly fallen snow will produce substrate luminances slightly below and above the ranges presented here.

CONCLUDING REMARKS

The picture presented in this chapter does perhaps at first sight seem confusing. Not only is the full range of night-time light level difficult for us to experience personally but night-time is also difficult to define. Even when night-time has been defined according to some arbitrary criterion it then becomes clear that what is experienced at night depends crucially on latitude and season.

Perhaps the most important message of this discussion is that we cannot simply extrapolate our own experience of night-time at one place and time to other places and times. For example, in Chapter 4 the occurrence of nocturnal migration in birds is discussed but it is important to understand that as a bird travels from a high arctic latitude towards wintering grounds in the equatorial regions it will experience a marked change in both day-length and in the minimum light levels of the night. A bird leaving latitude 76° at the end of the breeding season will not have experienced night-time light levels for many weeks. Much of its migratory flight to temperate latitudes, although perhaps lasting more than 24 hours, will still not take place at the kinds of night-time light levels which nocturnally active birds living in the tropics will experience every day of the year.

Figure 2.5   The luminance (log cd/metre^2) of a natural substrate, such as leaf litter or grass, when illuminated by various natural light sources. Along each vertical scale the full ranges of luminance levels between maximum sunlight and minimum starlight are shown. The range of luminance levels produced by the moon at various altitudes and phases, and by natural starlight, are also indicated. "Open habitats" refers to any situation where the substrate is not shaded by vegetation. "Closed habitats" refer to substrate luminance levels which would occur beneath a closed canopy woodland, which attenuates the incoming light by 2.0 log units (100-fold). Maximum cloud cover is considered to reduce light levels by 1 log unit (10-fold). Also shown is the approximate luminance of the sky near the horizon under the various natural conditions indicated. Data derived from the naturally occurring illumination levels shown in Figure 2.1 and Table 2.1, and measures of the average attenuation of a broad leaved, tree canopy in full leaf, by Federer and Tanner (1966), and converted to luminance levels assuming an average reflectance of the substrate of 25% (based upon the measures of Martin (1982)) compared with an "ideally reflecting and diffusing" surface as defined in the text. Data on the illuminance of the sky from Middleton (1958).

Perhaps the most important point to note from this discussion is that even for truly nocturnal birds in temperate or tropical latitudes the light levels of night-time are not uniform. It is best, therefore, when trying to understand the sensory problems which a nocturnal life style poses for an animal, to be aware that the animal will have evolved strategies to cope with a wide *range* of light levels and not any one arbitrarily defined "night-time" light level.

CHAPTER 3

Defining nocturnality

In the preceding chapter it was seen that a simple definition of night-time in terms of light levels is very difficult. Night-time, however defined, embraces a wide range of light levels. It is not surprising, therefore, to find that defining the "nocturnal habit" is also problematical. It is not only necessary to define night-time, but it is also necessary to consider which activities an animal should engage in during this time for it to be regarded as strictly nocturnal, as opposed to an animal which may show only occasional nocturnal activities.

Perhaps the most satisfactory definition of night-time in terms of light levels would be to consider it as embracing the full range of light levels which can occur naturally below that produced by maximum moonlight under cloud free conditions. Reference to Figure 2.5 shows that this would embrace a range of light levels in excess of 1.2 million-fold depending upon the presence of the moon and cloud cover and whether one was in an open habitat or beneath a thick woodland canopy. The case for taking the upper limit as the light level produced by maximum moonlight rests upon the fact that this is the maximum light level which *can* be experienced "in the middle of the night".

However, the light levels produced by maximum moonlight actually fall within the light level range experienced during the nautical twilight period (Tables 2.1 and 2.2). The question arises therefore as to when night-time, and hence nocturnal activity, should be considered to begin and end within each 24 hours of the daily cycle. Should it begin, for example, when light levels fall during the twilight period to those equivalent to maximum moonlight?

One requirement is that any definition of the nocturnal habit should be operationally easy to apply in the field where observations of behaviour may be made. Thus, choosing a particular light level [which will in any case occur at a time of rapidly changing light levels (Figure 2.1)] will not be easy to apply in the field. Also, choosing a particular time with respect to sunset is unsatisfactory because, as discussed in Chapter 2, the length of twilight and the associated changes in light level are functions of both latitude and season (Figures 2.3 and 2.4).

For these reasons it is proposed that the following simple definition has some utility: *a strictly nocturnal animal is one which habitually completes all waking activities of its life cycle when ambient light levels are below those produced in open habitats at sunset or sunrise.*

This definition has two principal advantages. First, setting the time of activity between sunset and sunrise provides reasonably rigid defining points which all observers could agree upon; and secondly, it excludes animals which may exhibit only occasional nocturnal activities during their life cycle.

There would, however, seem to be two principal disadvantages of such a definition. First, it implies that night-time begins at sunset and thus the maximum range of nocturnal light levels extends to approximately 1,000,000,000-fold (9 log units) compared to the one million-fold when maximum moonlight is used to define the upper limit. However, the minimum light levels which can naturally occur at sunset (extreme cloud cover under a vegetation canopy) are in fact approximately equal to those of maximum moonlight (Figure 2.5). Secondly, there are many animals, which would not be regarded as strictly nocturnal, but which are regularly active after the time of sunset and before the time of sunrise. For example, as will be discussed below, many otherwise diurnally active bird species occasionally forage at night. The key lies therefore in the idea that to be regarded as strictly nocturnal *all* activities of the animal's life cycle: foraging, finding a mate, reproduction, caring for young, etc, should be completed after sunset and before sunrise.

It should be noted however that while, in a book such as this, it may be important to make some attempt at defining the nocturnal habit, field observations of nocturnal behaviour and their subsequent translation into notes in handbooks and field guides, are unlikely to have been made with any systematic definition in mind, whether this is of the range of behaviours exhibited by the animal "at night", or of the light levels at which they were conducted.

Even when applying this rather broad definition of "nocturnality" it is found that the number of bird species which can be regarded as strictly nocturnal is very limited. Also, a converse category of *strict diurnality*, in which *an animal habitually completes all waking activities of its life cycle between sunrise and sunset*, is also found to be rather restricted. This is because such a definition will exclude all of the bird species which are at times regularly active after sunset, in some part of their life cycle. Clearly during the twilight period when light levels are changing rapidly a lot of otherwise diurnal birds of many different species may still be active.

The next chapter deals with those birds which are not strictly nocturnal according to the above definition but which nevertheless do on occasions engage in some purposeful activity outside of daylight hours. Some of these birds would often be regarded as strictly diurnal in that they usually roost throughout the night. Indeed many of these birds may typically enter roosts before sunset and stay there until after sunrise yet they may be active right through the middle of the night for part of their annual life cycle. It will be seen, however, that the nature of these occasional nocturnal activities is often very restricted.

Discussion of these activities and of how they might be restricted, and permitted, by sensory considerations will be of considerable interest when discussing just what the strictly nocturnal birds are capable of achieving at night. For convenience these birds will be considered under two main headings.

First, there are birds which may be considered under a heading of *occasionally nocturnal species*. These are birds which may complete a specific aspect of their life cycle during the hours of darkness. In none of these species does the behaviour completed at night seem to be exclusively nocturnal. In many cases it is an extension beyond dusk of an activity carried out during the previous and following day. In other cases occasional nocturnal activities are strictly seasonal in their occurrence. It cannot be concluded that these activities will occur at all of the nocturnal light levels described above. As will be discussed below many of these occasionally nocturnal activities seem to take place only in open habitat types and possibly also at the higher nocturnal light levels. Whether the behaviour is performed nocturnally may depend on factors such as the weather or the tidal cycle. In addition such nocturnal activity cannot always be regarded as a property of all individuals within a particular species. Whether an individual bird is occasionally active at night may depend upon many factors including, for example, its geographical location, rather than just the species to which it belongs.

Mention must also be made here of *crepuscularly active species*, i.e. species which are habitually or occasionally active around dusk and/or dawn. Although such activity is often recorded in handbooks and field guides it is not clear whether there are species or individuals which can be satisfactorily placed either exclusively or principally within this category. Twilight marks the transition between night- and day-time and at low latitudes this can be relatively brief while at higher latitudes it can be very long (Figures 2.3 and

2.4). In addition, providing a satisfactory definition of the light levels which occur at this time is inevitably problematical. Also it would seem that many principally diurnally-active species may on occasion be active before sunrise or after sunset, while nocturnal species may begin and end their periods of activity at some time during twilight, and this is in fact taken into account by the definition of nocturnality proposed above. It can also be seen from Figure 2.5 that the light levels experienced at noon on an overcast day, within a woodland, can actually be below those experienced at sunset in an open habitat; in other words, "twilight" light levels can actually occur in the middle of the day depending upon location and particular circumstances. Although there is a problem in defining crepuscular light levels some general points concerning activity during twilight can be made and these will be discussed in more detail in Chapter 5.

CHAPTER 4

Occasional nocturnal activities in birds

Most bird species begin roosting when light levels start to decrease to those experienced around dusk. The specific role of light levels in the initiation of such roosting seems not to be well understood (Krantz and Gauthreaux 1975; Amlaner and Ball 1983; Bacon 1985, p. 517). Indeed while there is some evidence that light level may provide the main factor which triggers roosting, or the daily onset of activity, precise timing may be controlled by a complex of factors both external and internal to the bird. These may include temperature and humidity (Astrom 1976), and endogenous rhythms of wakefulness and sleep (Gwinner 1975).

Among these typically diurnal birds are found many species which, under certain circumstances, exhibit occasional nocturnal activity. Such activities fall under four main headings, the most prominent of which are *night migration* and *night singing.* These two behaviours often occur in the same species and it seems likely that the majority, if not all, of the occasional night singers are also night migrants, though the reverse of this relationship is certainly not true. There is also a group of birds which may feed after dusk, but their *night foraging* is not an exclusively nocturnal activity and so it is most appropriately considered as an occasional nocturnal activity. Finally,

33

there is a group of birds which *arrive at and depart from their nest sites* during darkness but which appear to be principally diurnal during the rest of their life cycle.

NIGHT MIGRATION

There is good field evidence that many birds undertake at least part of their migratory journeys at night. This comes from many sources: the calls of birds passing overhead at night; direct observation of birds passing through bright light beams or across the face of the moon (e.g. the studies of Able and Gauthreaux 1975; Bingman *et al* 1982); the attraction (often fatal) of birds during night-time to lighted structures such as light houses, office towers, oil platforms and transmitter masts (Aldrich *et al* 1966; Durman 1976; Verheijn 1981; Bourne 1982); the observation of bird movements by radar techniques (there are many studies in this field, for example, those of Richardson 1976, 1978, and the summaries of Eastwood 1967, Richardson 1979, and Alerstam 1985), and the presence or arrival at dawn of birds at migratory watchpoints, such as bird observatories (Durman 1976).

There has also been much field and laboratory research aimed at understanding how birds might orientate their migratory flights. All of this work has led to general statements as to the prevalence of night migration among birds. For example, "Most songbirds (order Passeriformes) migrate almost exclusively at night" (Able 1982, p. 761) or "Most long-distance passerine migrants, travelling from temperate or Arctic breeding areas to tropical non-breeding areas, depart an hour or two after sunset" . . . "most short-distance passerine migrants travelling chiefly within the northern temperate zone, fly by day" (Evans 1985, p. 349). Despite these statements systematic lists of species which are known to make at least some of their migrations at night are not available. Even the question of whether bird species fall neatly into "nocturnal" or "diurnal" migrant categories, as is implied by the discussions in many radar studies of migration, seems not to have been widely discussed. Indeed Curry-Lindahl (1981) devoted only four pages, out of a total of nearly 700, in his book on *Bird Migration in Africa* to the topic of migration at night. In those four pages he paints a very confused picture and calls for more systematic observations to be made. Also, recent reviews and multi-author works on migration (e.g. Schmidt-Koenig and Keeton 1978; Gauthreaux 1980; Baker 1984; Mead 1983) fail to discuss the actual occurrence of night migration in a systematic way.

Clearly, night migration is a phenomenon which every ornithologist knows about but one which is poorly documented and understood. However, in terms both of the total number of species and the number of individuals involved, night migration is probably by far the most extensive nocturnal behaviour that birds exhibit. So what does the phenomenon of nocturnal migration consist of?

First, it would seem safe to suggest that most nocturnal migration involves species which are diurnally active during the rest of their life cycle, i.e. these

Fieldfares *Turdus pilaris*, like other thrushes, migrate by both night and day.

birds become nocturnally active only when they are migrating (some species, however, may also occasionally forage or sing at night; see the sections on occasional night foraging and singing below). Studies of circadian and circannual rhythms in birds (mainly in passerine species held in cages), have shown that at the time of year when migration usually takes place, instead of roosting, these birds begin to show activity after the end of the normal daylight period. This activity has become known as "migratory restlessness" and in caged birds is usually recorded by monitoring how often the birds hop between perches, for example the work of Gwinner and co-workers (Gwinner 1975, 1985; Gwinner *et al* 1985). It has also been shown in the European Robin that individuals may even exhibit a preference for lower light levels for this activity during the normal night-time period (Gwinner 1975). At all other times of the year these birds begin to roost at the end of the daylight period.

Secondly, there is evidence that the categories of night-time and day-time migrants are not mutually exclusive at either the species, population or individual level. For example, within one species there is evidence that some individuals migrate by day and others migrate at night, even through the same migratory watch points, as in the European Starling *Sturnus vulgaris* (Feare 1984; Evans 1985, p. 349). Also, species which by general consensus are regarded as day-time migrants, such as the Swallow *Hirundo rustica,* Swift *Apus apus* or Chaffinch *Fringilla coelebs,* may occasionally arrive at migratory watch points during the night (Durman 1976; C. Mead, personal communication 1988).

Thirdly, there are species in which the majority of birds regularly migrate continuously by both night and day, at least for part of their migratory journey. This is exemplified among the passerines by species which undertake particularly long distance migrations over water. For example, the Blackpoll Warbler *Dendroica striata* and the Wheatear *Oenanthe oenanthe* contain populations which are thought to make non-stop journeys of more than 24 hours duration over the western and eastern Atlantic Oceans respectively (Snow 1953; Dorst 1961; Nisbet 1970). It is also believed that many shore-birds and wildfowl which regularly migrate between high Arctic latitudes (Greenland, Arctic Canada, the Svalbard Archipelago) and Europe, must do so on continuous journeys lasting more than 24 hours (Evans 1990). However, it should be pointed out that, at the time of year when these journeys take place, night-time may not occur or at least be of only short duration, at high latitudes (Chapter 2). Hence these long journeys may not necessarily entail nocturnal flights.

Fourthly, having departed at or soon after dusk, nocturnal migrants may sometimes complete a section of their journey before, rather than after, dawn. Field observations show that birds which arrive at their destination during the night typically remain exactly where they land. They only begin to move around and seek cover and food at about dawn. Thuc C. Mead (personal communication) reports finding unidentified warbler species roosting before dawn on an open beach on the south coast of England, and P. Berthold (personal communication) reports that Passerines arriving during the night in the vicinity of the ringing station at Radolfzell on the shores of the Boden-see in southern Germany, do not appear to move into surrounding scrub and trees until some time around dawn.

Fifthly, radar studies have shown that night migrating birds are flying at considerable altitude, typically above 500 m but often much higher than this. Passerines are often above 2,000 m and have even been recorded as high as 6,800 m (Lack 1960; Richardson 1976). During their nocturnal flight the birds may change altitude in a systematic way but they still remain well clear of the ground or sea surface (Bourne 1982).

Few authors have considered the question: "Why migrate at night"? Evans (1985, p. 349) proposed that there were several "advantages" for migration at night: the availability of the fixed pattern of stars as a navigational cue; reduced likelihood of predation; improved atmospheric conditions for the detection of sound signals from the ground or from other migrating birds; the fact that daylight hours become available for feeding. More recently Kerlinger and Moore (1989) have proposed that the structure of the atmosphere at night may be more favourable for flight compared with midday. Their proposal is discussed in more detail below. Some, or all, of these factors may be important in determining whether birds migrate by night or by day. However, a more thorough understanding of the actual species involved in night migration, and their specific patterns of movements, would seem to be necessary before any general "advantages" of night migration can be determined.

WHICH SPECIES MIGRATE AT NIGHT?

In view of the information presented above it is worth asking the simple question, "Which species migrate at night?" Some studies, for example, Kerlinger and Moore (1989), bring together diverse sources of information about the prevalence of nocturnal migration among certain groups of birds classified according to taxonomic order. These data show much disagreement between studies. However, some of this disagreement could result from differences in the methods of study as well as marked differences in the geographical locations of the sites where data were obtained. In view of this, and the lack of published systematic lists of the nocturnal migratory behaviour of species passing through the British Isles, a simple survey was conducted.

This survey collated data on the normal patterns of nocturnal and diurnal migratory behaviour through seven different bird observatories around the coast of the British Isles. Data were obtained by asking observers to record what they regarded as the normal pattern of nocturnal–diurnal migratory behaviour of the bird species which pass through each observatory. Each observer was either based permanently at the observatory or had spent a great deal of time there during the migration seasons and had access to daily schedules of birds recorded.

Appendix 1 shows examples of the kind of data obtained for different bird species. For some species groups included in this survey, such as the finches and warblers (Fringillidae and Sylviidae), there is reasonable agreement between observatories as to whether the birds are principally nocturnal or diurnal migrants. However, for other groups such as the waders (Charadriiformes) reports from the observatories showed marked differences. For example, the Dunlin *Calidris alpina* is regarded by different observers both as principally nocturnal and diurnal as well as falling into the various intermediate categories.

Appendix 2 shows the species listed according to a "final category" (see Table caption) which summarises the reports from the individual observatories for all the 147 species for which data were available. It is clear that these tables could be analysed from many different points of view. For example, can systematic patterns of nocturnal–diurnal migration be discerned which can be accounted for in terms of the taxonomy, ecology, or geographical distribution of the species? Before discussing these questions, however, two preliminary general points arising from these Tables do seem particularly pertinent to any discussion of nocturnal migration.

First, in view of the good agreement between observatories over some species, it seems likely that where differences do occur in the reports of the different observatories (for example as in the Charadriiformes in Appendix 1), they reflect genuine differences in the degree to which these birds may migrate at night as they pass through various observation points around the British Isles.

Secondly, in the summarised data of Appendix 2 it can be seen that contrary to expectations no species fell into the exclusively nocturnal migration category, i.e. no species was recorded as a nocturnal migrant at all observatory watch points. Even species falling into the mainly nocturnal migration category were relatively rare (29% of passerines and 5% of non-passerines). However, equally nocturnal and diurnal migratory activity was commonly reported both among passerines and non-passerines, but nearly half of all passerines and two-thirds of all non-passerines are either diurnal or principally diurnal in their activity through these particular migratory watch points. Even so, 75% of the species have been recorded as a night migrant at some time. Thus, while many birds which would commonly be acknowledged as diurnal during most of their life cycle do migrate at night, it cannot be concluded that any species *always* migrates at night through these particular watch points. However, many otherwise diurnally active species are capable of travelling at night, depending upon the circumstances.

Just what these circumstances might be is, however, open to debate. The actual patterns between species of migratory activity revealed in Appendix 2 are very intriguing but understanding them poses many problems. It would be possible to attempt to classify these birds according to many criteria (e.g. taxonomic classification, geographical breeding and non-breeding areas, length of migratory journey, migratory route, diet and foraging techniques, size of bird, wing structure and speed of flight) and attempt to find a factor or factors which correlate strongly with the propensity to migrate at night.

The migratory behaviour of some species groups does seem readily amenable to explanation, that of others does not. For example, those birds which generally appear unwilling to undertake sustained powered flight for long periods but, rather, use soaring flight (e.g. the broad winged raptors, Accipitriformes and Falconiformes) migrate by day, presumably in order to make use of thermal lift which is reliably generated during day-time (Pennycuick 1969). However, it is unlikely that this is a factor which can account for the diurnal migrations of the other species so listed (e.g. the divers (Gaviiformes), Gannet and cormorants (Pelecaniformes), swallows and martins (Hirundinidae)) since these include birds known to be capable of sustained powered flight. Furthermore, there are reports that broad-winged raptors are in fact capable of sustained powered flight (Kerlinger and Gauthreaux 1985a,b).

Length of journey, migratory route and size of bird would also seem insufficient as general explanatory ideas for the pattern of nocturnal migration seen through British Bird Observatories. It may be seen, for example, that some groups of birds which are similar in these respects [such as the hirundines, pipits (Motacillidae) and the warblers (Sylviidae)], may differ in their propensity towards nocturnal and diurnal migratory flight through British observatories.

It would seem that no single variable of those discussed above offers a satisfactory *overall* account of the migratory pattern seen in Appendix 2. In the light of this, two general hypotheses which may throw some light on the general pattern of nocturnal–diurnal migratory activity can be suggested.

First, one of the most obvious correlations revealed in Appendix 2 is between the migratory categories and taxonomic classification, principally at the family level. Thus, for example, the Turdidae (thrushes) and the Sylviidae (warblers) would seem to be more likely to travel at night to and from the British Isles compared with the Fringillidae (finches) and the Motacillidae (pipits and wagtails). There is a case, therefore, for proposing that these family groups differ physiologically, perhaps in their sensory capacities or in their ability to make use of the orientational cues which are available at night. To consider this possibility in more detail requires an analysis of the sensory capacities of birds and the cues which night- and day-time migrating birds might use. These are discussed briefly below but it should be noted, now, that there is no relevant information on interspecific differences in the sensory or perceptual capacities of these birds.

The second suggestion is that the interspecific pattern of nocturnal–diurnal migration in Appendices 1 and 2 can be at least partly accounted for by the idea that individual birds or species are nocturnal migrants only under certain meteorological conditions, or only along certain sections of their migratory route. If this idea has any validity then whether a bird is reported as a nocturnal migrant may depend upon aspects of the night environment and also upon the point at which its behaviour is sampled along the migratory route. Thus, nocturnal migration becomes a "local" strategy which is adopted by a bird to cope with conditions along the route rather than an overall strategy which certain individuals or species adopt for the entire length of their journey.

Cramp (1988, p. 598) describes the European Robin (*Erithacus rubecula*) as, "A nocturnal migrant, though local movements occur by day". This distinction between "migratory" and "local" movements may be of little functional significance to a bird that is actually migrating. It may simply reflect the pattern of nocturnal and diurnal migration adopted at different points along its route and/or from day-to-day, according to some aspect of the night environment or geographical barrier. It could also reflect the difference between actual nocturnal flights and the so-called "redetermined migration flights" which apparently take place by day (Gauthreaux 1978), although whether they are exclusively diurnal is not known. (Such redetermined migration flights are thought to occur when a bird strays from its intended migratory route.)

NOCTURNAL MIGRATION AS A "LOCAL" STRATEGY

Crossing inhospitable terrain

One important factor which may determine whether a bird migrates at night may be the length of time necessary to fly across continuous inhospitable terrain on the route adopted by the population to which it belongs. In support of this suggestion are the following observations based in part upon the data of Appendix 2.

(1) The diurnally migrating seabirds (divers, shearwaters, gannet, gulls, terns and skuas) in one sense never have to cross inhospitable terrain. They are capable of coping with most sea conditions and can therefore halt their journey before nightfall and then continue the next day. More importantly perhaps, they are usually capable of finding feeding grounds at many places along their migratory routes. Thus at no point do these mainly diurnal migrants pass over inhospitable terrain which *must* be, or is best, covered without stopping.

This does not apply so readily to either the shorebirds (Charadriiformes) or the wildfowl (Anseriformes), for although these birds can swim, the majority of them cannot feed out at sea (indeed suitable feeding areas may be relatively far apart) and neither can they cope with inclement weather as readily as true seabirds. Thus for these birds any long sea-crossing presents a terrain no less inhospitable than it would be for land birds making the same journey. Although the flight speed of shorebirds and wildfowl is often relatively high compared with passerine species (Rayner 1985), it will often not be possible to complete sections of a migratory journey within less than 24 hours and hence nocturnal flight will be necessary.

(2) Although the migratory pattern amongst the other non-passerine species and the passerine species is not quite so straightforward, the same principle of nocturnal migration being a local rather than a universal strategy, would still seem to apply to most species as listed in Appendix 2.

Thus, if a migratory route involves the crossing of a large tract of inhospitable terrain where feeding is not possible (such as sea, desert or mountains) and if the bird's flight speed is such that the barrier cannot be crossed in a non-stop flight taking less than the available daylight hours, then at least occasional nocturnal migratory activity would be expected. If such a barrier is present, then the more successful strategy might be to feed all of the day prior to departure at dusk across the barrier. Departure at dawn on the same migratory journey would preclude the possibility of being able to feed right up until departure and could, perhaps, entail arriving at a safe haven on the other side of the inhospitable terrain during the night when again feeding would not be possible. However, some barriers of inhospitable terrain are so extensive, and flight speed is so low, that migratory journeys must continue for more than 24 hours, thus making nocturnal flights essential.

In the context of the Palearctic-African migratory system two major inhospitable barriers, the Sahara Desert and the Mediterranean Sea, confront birds which travel to Europe from south of the equator. Although it is believed that many birds fly on routes which skirt the edges of these barriers (for example, Hilgerloh (1989) has shown that many southward migrating passerines travel over the eastern Atlantic Ocean and thus avoid the western Sahara) some birds are known to cross the central parts of the Sahara and the Mediterranean Sea. North to south the Sahara extends about 2,000 km and, taking the cruising flight speed of small passerines to be in the region of 30–

40 km/h (Rayner 1985), it can be seen that these birds must require between 50 and 70 hours of flight time to cross this barrier.

The Yellow Wagtail (*Motacilla flava*) presents a particularly well-studied and instructive example of a trans-Saharan migrant. These birds may be nocturnal only on those sections of their migratory route which cross inhospitable terrain. Certain populations of Yellow Wagtail breed in Europe and winter south of the Sahara (Cramp 1988, p. 413). It has been estimated that prior to the spring migration these birds deposit enough fat for a 60–70 hour continuous flight and that they probably cross the desert in a non-stop flight lasting three nights and two days (Wood 1982). By departing near to dusk the birds should, given favourable weather conditions, arrive north of the Sahara at around dawn two and a half days later. Biebach *et al* (1986) have also presented evidence that, in autumn, Yellow Wagtails crossing the Sahara travel at night and during the early morning, but may rest at oases or in the shade of rocks during the middle of the day. Such a pattern may be necessary to reduce evaporative water loss which would be lower when flying in the cooler night-time than during the high temperatures of the day. It may even be that this species adopts this strategy only when crossing the Sahara in the autumn, whereas in spring on the northward journey the birds travel continuously by day and night (Biebach, personal communication).

Such flights may in fact be typical of many trans-Saharan migrant passerines but the case of the Yellow Wagtail is particularly interesting since it is recorded as a mainly diurnal migrant at many observatories around the coast of the British Isles (Appendix 2) and also in reference works (e.g. Cramp 1988, p. 418). Thus it seems possible that this bird migrates at night in order to cross a major inhospitable barrier but perhaps travels via a series of shorter stages during the day on other sections of its journey. The Hirundines and other Motacillidae (pipits and wagtails) perhaps parallel the Yellow Wagtail in that they may cross the central Sahara by night but are generally regarded as day-time migrants at British Observatories.

Migrations within Europe often involve crossing the North or Baltic Seas or the English Channel. All of these vary in their width but all are sufficiently wide to require a long continuous flight if crossed at other than their narrowest points. The minimum width between Scandinavia and the British Isles is approximately 650 km which is sufficient to necessitate a nocturnal flight by the passerines and shorebirds.

Although the above scheme may provide a satisfactory framework for explaining most of the general pattern shown in Appendix 2 it must be stressed that each species, or even population and age class within that species, really requires individual examination, especially among the species recorded as both nocturnal and diurnal migrants. It must also be stressed that the particular pattern of diurnal/nocturnal migration seen in Appendix 2 relates to observations from around the coast of the British Isles which may be influenced by sea crossings. These same patterns may not apply to the same species in other parts of Europe. However, the example of the Yellow Wagtail does suggest that the propensity towards nocturnal migratory behaviour may change along the route.

Local atmospheric conditions

Kerlinger and Moore (1989) have argued that local atmospheric conditions along the migratory route may also favour migration at night, rather than by day. This is because under certain conditions flight at night may be faster and require less energy. Kerlinger and Moore have not been able to quantify these general energetic benefits of night migration which may arise due to three principal factors. (1) At night, air temperatures are lower and relative humidity is often higher than by day. These serve to reduce evaporative water loss from the bird's body. In addition the cooler night air is denser. Therefore generating lift in powered flight is energetically less costly in this cooler air. (2) Horizontal winds at midnight tend to be slower than at midday so birds are not as likely to be blown off course at night. (3) As a result of decreased thermally generated turbulence, winds at night are less variable in direction than winds at midday. This means that night migrating birds may need to change heading and air speed less often than during the day in order to maintain a straight and level flight path over the ground. This is also likely to reduce the total energetic cost of the migratory journey.

Thus it may be that individual birds may decide to travel by night rather than by day on the basis of local atmospheric conditions and possibly also according to their own energy reserves for the flight.

HOW DO NOCTURNAL MIGRANTS GUIDE THEMSELVES?

For the person interested in nocturnal behaviour the above discussion throws up many intriguing problems. Although the actual species involved and their patterns of nocturnal/diurnal migratory behaviour may not be well understood it is clear that a large number of otherwise diurnally active species do migrate at night on at least some part of their journey. It may well be that nocturnal migration is not the preferred strategy. Rather it may be born from the necessity of crossing an extensive tract of inhospitable terrain, or the seeking of more favourable atmospheric conditions for long distance flight. Nevertheless the vast amount of data from ringing studies and the innumerable field observations of amateur and professional ornithologists all attest to the fact that nocturnal migration is a successful strategy for the majority of birds which do chose to travel at night. The question arises therefore as to how birds which normally roost at dusk are able to fly and navigate successfully at night-time for a brief period of their annual life cycle? Normally these birds seem ill-equipped to be abroad at night, and this presumably stems from limitations imposed upon their behaviour by their senses. So what is the sensory basis of their nocturnal migratory flights? As mentioned above, do species differ in their sensory capacities or in their ability to make use of the orientational cues which are available at night?

A great deal has been learnt from laboratory and field experiments of the possible sensory cues that birds migrating at night might use to determine the *direction* in which to travel, at least at the beginning of their journeys.

However, such studies deal literally with only a handful of passerine species drawn from just four families, the warblers (Sylviidae), flycatchers (Muscicapidae), buntings (Emberizidae) and thrushes (Turdidae). The cues so far identified include the position of the setting sun, the earth's magnetic field, the patterns of stars, the moon and topographical landmarks. It would be inappropriate to discuss all of the evidence concerning the use of these various cues in different species, but this field has been reviewed a number of times [see Able (1980), Papi and Wallraff (1982), Baker (1984), Presti (1985), Rankin (1985), Able and Bingman (1987), Moore (1987) and Wiltschko and Wiltschko (1988a,b) for reviews which emphasise different aspects of the role of these cues in the navigation of birds]. The interpretation of some of this evidence is still controversial, especially that concerning the use of the earth's magnetic field as a compass and its interaction with other cues (Wiltschko and Wiltschko 1988a,b), and it is perhaps sufficient to note here that a number of cues may be available simultaneously to a migrating bird, though which cue may be used by a given species is not clear. It would certainly be misleading to assume that all of the cues are equally available to all birds when migrating at night.

There is evidence that whatever cues are available they are used in some kind of hierarchical or redundant fashion, such that one cue can substitute for another in differing circumstances (Emlen 1975, Wiltschko and Wiltschko 1988a). This means that one species of population of birds (or even age class within a species) in one set of weather conditions, or geographical locality, may rely primarily on one set of cues to determine the orientation of their nocturnal migratory journey, while another group of birds may use another set of cues (see the above-mentioned reviews and, for example, the results of Able 1982 and Sandberg *et al* 1988).

Visual cues

The importance of the earth's magnetic field in the migratory or homing navigation of at least some bird species seems generally accepted (Baker 1984; Presti 1985; Wiltschko and Wiltschko 1988a,b). [Evidence for magnetic field detection has so far been presented for the following Passeriform and other birds: European Robin (*Erithacus rubecula*), Whitethroat (*Sylvia communis*), Garden Warbler (*S. borin*), Subalpine Warbler (*S. cantillans*), Blackcap (*S. atricapilla*), Indigo Bunting (*Passerina cyanea*), Savannah Sparrow (*Passerculus sandwichensis*), Pied Flycatcher (*Ficedula hypoleuca*), the Pigeon (*Columba livia*) and the Ring-billed Gull (*Larus delawarensis*).]

However, while magnetic cues from the earth may be available to these birds, the main body of evidence suggests that cues based upon vision are of equal or greater importance to magnetic cues in the orientation of nocturnally migrating birds. At least, if magnetic field cues are used, they are thought to be employed to calibrate a visually based cue rather than the magnetic field be consulted directly. In fact, although the sensory basis of magnetic field detection is not clearly understood (Walcott and Walcott 1982), there is growing evidence of a possible intimate relationship between the mechanism of magnetic field detection and the visual system [Leask (1977); Semm *et al* 1984; Semm and Demaine 1986].

The following experimental findings and field observations attest to the importance of visual cues in the guidance of nocturnally migrating birds, although it must be remembered that many of these findings strictly apply to only a handful of species.

First, the majority of nocturnal migrations take place in weather which provides the "ideal" conditions of calm, light or following winds, with little cloud cover and good visibility both prior to the time of departure and during the actual flight (Richardson 1978; Elkins 1983).

Secondly, before their nocturnal migratory journey passerine birds usually cease their normal day-time activities and start to roost around dusk in the usual way (Palmgren 1944; Hebrard 1971). They then become active again and typically depart on their migratory flight between 30 and 45 minutes after sunset, usually after the time of civil twilight (Parslow 1969; Richardson 1972; Hebrard 1971). It is thought that it is during this pre-departure period that migrant birds make the decision of whether to migrate and in which direction. Just what birds do at this time is unclear, but the primary importance of visual cues associated with detecting the position of the setting sun or the pattern of polarized light in the sky (which is a direct correlate of the sun's position) has been demonstrated (Able 1982; Able and Cherry 1986; Moore 1987). The extent to which a bird's migratory orientation is influenced at this time by the earth's magnetic field is unclear. If magnetic cues are employed they may be used to calibrate a compass based upon the stars rather than consulted directly, but evidence is conflicting (Wiltschko and Wiltschko 1975a,b; 1976; Sandberg *et al* 1988).

Thirdly, there is considerable evidence that visual cues are of primary importance not only in determining the initial orientation of migratory flight but also in maintaining it throughout the nocturnal journey. These visual cues may be associated with the stars and moon and also involve the use of topographical landmarks as guides or beacons.

That caged nocturnal passerine migrants (Blackcap, Garden Warbler and Lesser Whitethroat) may orient themselves using the pattern of the stars above, was first demonstrated in experimental cages by Sauer and Sauer in 1955. Since that time these results have been refined and extended to a few more species especially the Indigo Bunting *Passerina cyanea* (Emberizidae) (Emlen 1967). However, the extent to which star patterns are used to orient free-flying birds during their migratory flights is less clear. Evidence from both radar studies (Able 1982) and direct visual observation (Able 1974) has shown that although birds prefer to make their migratory flights under clear skies, those which do fly under overcast skies may be no less well oriented in their flight path than those which travel when stars are visible.

Nocturnal migratory birds seem to become disoriented in flight only when both the ground and the stars are not visible to them. Thus disorientation among migrating passerines is very rare (Bellrose and Graber 1963; Drury and Nisbet 1964; Bellrose 1967; Able 1974) and seems to occur only when birds are deprived of visual cues. So long as birds fly below the cloud ceiling, their flight is, "as straight, level and fast as comparable birds flying on clear nights" (Able 1982, p. 47). Cases of poor orientation are associated with low overcast, rain, fog of cloud cover, often of several days duration (Emlen 1975), although there are contested reports that birds can still be well oriented even under these conditions (Griffin 1973). In addition Able (1982) has argued that simple absence of access to sun and stars for several days need not result in disorientation of nocturnal migrants. "So long as birds were flying below the clouds, tracks were as straight as under clear skies and the distribution of mean headings of all birds was oriented" (Able 1982, p. 48).

The disorientation of birds in cloud or fog does suggest that they cannot navigate successfully if reliant on cues from the earth's magnetic field alone. Some kind of visual cues, derived either from the stars and/or moon above or from the ground below, would seem necessary for the correct orientation of birds at night. The fact that migrating birds are attracted, often fatally, to isolated illuminated structures suggests strongly that at night these birds may be dominated by visual cues from below them but that they are easily confused.

Evidence which throws light on the basis of this confusion is rather limited. There are observations which suggest that nocturnally migrating birds do not even need to be caught out by "bad weather" for them to be confused by artificial lights below them. D. Pearson (personal communication 1988) has regularly observed that passerine species migrating through East Africa are attracted at night to floodlights surrounding water holes but only under certain conditions. There must be fog or mist at ground level in the area illuminated by the lights and no moon. When these two conditions are fulfilled birds passing even beneath a clear, starlit sky but above the fog or mist layer are attracted through it to the lights beneath. If a moon (of any phase) is present then these lights are ignored, as also they are on clear nights without ground level fog or mist. Pearson even reports that on some occasions when birds were attracted to the lights he could still see stars through the layer of mist near the ground.

These observations are paralleled by data collected at Bardsey Island bird observatory (Wales) (Durman 1976). Here there is a strong correlation between the number of birds attracted to the lighthouse and the absence of the moon (even to the extent that birds were attracted temporarily to the light during an eclipse). Here also birds may be attracted to the light when there is low level cloud or "smoke haze" and a clear starlit sky above.

It would seem that to attract the birds the light must be scattered by fog or mist to produce an isolated pool of diffuse light. In the case of lighthouses with revolving lights the birds even appear "trapped" by the moving beam and try to remain within it. Presumably the rotating beam provides a constantly present pool of scattered light. However, if the lighthouse employs a flashing light which does not rotate, and is thus simply on or off, the birds seem little attracted, even under conditions of low level mist (C. Bibby, personal communication 1989). If an isolated lighthouse is surrounded by flood lighting, birds continue to be attracted to the area under the no moon, low cloud/haze conditions but are less likely to be fatally attracted to the lamp (S. McMinn, S. Cowdy, personal communications 1988).

Verheijn (1980) and Baker (1984, p. 94) have discussed the idea that the attraction of birds to lighthouses and other illuminated structures may be due to their mistaking the artificial light for the moon and that they would normally use the moon in the orientation of their flight. A further possibility is that the absence of the moon coupled with low cloud or mist may obscure the horizon. This is because under these conditions the contrast between ground and sky is greatly diminished. Aircraft pilots find the absence of an horizon particularly difficult for night flying and indeed to fly under such conditions requires special training and often a reliance upon instruments rather than visual cues from outside the aircraft (Pennycuick, personal communication 1989). It has been argued from theoretical considerations that perception of the horizon is essential for both day and night navigation in birds (Pennycuick 1960), while Whiten (1978) has presented experimental evidence that the Pigeon may require perception of an horizon when navigating.

Whatever the basis of these perceptual confusions they do suggest that these passerine birds' ability to make visual discriminations at night is rather limited and that the birds are influenced by visual cues from the ground beneath them even when there is a clear starlit sky above. Thus it may be suggested that visual cues from below may dominate over magnetic and star based cues during the actual migratory flight even though these cues may be used to determine direction of flight at the time of departure.

That day-flying migrating birds are typically influenced by visual cues from the landscape beneath them seems well accepted, together with the idea that these birds frequently follow so called "leading lines". These are usually large topographical features such as rivers, hill and mountain ridges, coasts and valleys (Mead 1983; Baker 1984; Evans 1985). It has also been shown that birds crossing extensive tracts of water can compensate for the drifting effects of wind by using the wave pattern below them to maintain a constant heading (Alerstam and Pettersson 1976).

Evidence from radar studies suggests that nocturnally migrating birds may also be influenced by similar topographical features, whether these are river systems (Bingman *et al* 1982) or mountain ranges, valleys or individual peaks (Bruderer 1978; 1982) although Emlen and Demong (1978) have argued from a study of the White-throated Sparrow (*Zonotrichia albicollis*) that the influence of topographical cues may be minimal for the orientation of these birds at night. The principal idea behind the use of landmarks in migrating birds is that the birds may determine their compass direction from some other cue but that this direction is then projected onto the landscape below. Topographical features which correspond with this direction can be noted and used for guidance. These may be in the form of leading lines, a single fixed beacon or involve the alignment of two or more distant objects. A single fixed beacon can act as a good guide but the drifting effects of the wind may mean that the beacon is approached along a "creeping curve" rather than directly. The alignment of two or more distant objects can produce a more direct course since it is then possible to make use of the effect of parallax (Bruderer 1982). As is discussed below, it may even be possible for birds to fly on a course aligned by distant features ahead and behind. This too would facilitate flying on a straight course even when the wind was liable to cause drift.

Baker (1978; 1984), in his model of bird migratory behaviour, suggests that on its first migratory journey a bird builds up a "mosaic map" of its migratory route. In this map, knowledge of suitable staging areas are linked together by compass orientations of flight paths and/or knowledge of topographical features of the kind described above. On subsequent migrations the bird may need do no more than consult its cognitive map in order to travel between breeding and non-breeding areas. Under these conditions its journey would then be composed of stages guided by known topographical features. According to this model, whether a bird is on its first (exploratory) migratory journey, or is an experienced adult, can potentially make a considerable difference to its degree of reliance upon compass and topographical cues regardless of whether it is migrating by day or by night. Baker (1984) argues that the experienced bird could complete its migratory journey guided by the now familiar topographical (visual) cues alone.

Visual fields

The idea that a bird projects the compass direction for its migratory flight on to the landscape is very similar to what humans might do when orienteering through unfamiliar terrain. However, the bird has two particular advantages over a person in this task. Not only can it gain altitude rapidly and thereby considerably broaden the range of topographical features brought into play, but the extent of its visual field will possibly allow the spatial relationships between different topographical features to be more easily appreciated.

By virtue of the optical structure of the eyes and of their position and movements in the skull, many birds probably gain complete coverage of the visual world both above and around them. Although the Woodcock *Scolopax rusticola* is often cited as an unusual bird in that it has completely panoramic vision it has recently been shown that other less notable birds may also have equally extensive visual coverage. Indeed in the Mallard Duck *Anas platyrhynchos* total panoramic vision, including complete coverage of the celestial hemisphere above the head, is achieved without eye movements (Martin 1986b), while in the European Starling a similar degree of visual coverage is possible by virtue of eye movements (Martin 1986c). Thus while a bird with laterally placed eyes flies towards or over a given topographical feature, the position of these features can be constantly monitored with respect to other features to the side and *behind,* and also to the complete celestial hemisphere above the bird. On a cloudless night this will mean that the star pattern of the whole hemisphere is constantly available and can be related to the complete landscape surrounding the bird. Clearly this could make the use of topographical cues for the guidance of the flight an easier task than it would be for ourselves even if we had the advantage of height that can be gained when

The Mallard *Anas platyrhynchos*, like many other birds with eyes placed laterally in the skull, has almost total visual coverage of the world about it.

flying. Our view of the world is dominated by a rather narrow forward view which takes little cognizance of the sky above. This view perhaps prevents us readily appreciating the spatial relationships between ourselves and all that is around and above us by day or by night. For the majority of birds such relationships would be far more readily apparent.

NOCTURNAL MIGRATION IN PERSPECTIVE

This section on nocturnal migration has raised a number of interesting points about the general nature of occasional nocturnal activity in birds. Discussion of these will provide a perspective for viewing other occasional nocturnal behaviours and for discussing the special features of strict nocturnality.

First, it has been shown that truly nocturnal behaviour, i.e. behaviour occuring "in the middle of the night" well away from the twilight periods of dusk and dawn, does occur commonly as an occasional activity in otherwise diurnal birds. However, it cannot be assumed that such occasional nocturnal behaviour is a feature shared by all members of a species. In addition there is no information on how light levels at night influence the migratory behaviour of individual birds or populations of birds.

Secondly, it cannot be assumed that nocturnal migration is a preferred behavioural strategy. The finding that most nocturnal migrants may also travel by day (Appendix 2), and indeed the suggestion that there are no "exclusively nocturnal" migratory species, suggests that there is no general advantage in migrating at night. On the contrary, it could be argued that the advantage may well lie in travelling by day and it is only of necessity, under

specific local circumstances, that birds travel at night. The two most important factors here may be the local geographical conditions (inhospitable
terrain), and local atmospheric conditions.

Thirdly, although night migrating birds clearly have at their disposal a
range of mechanisms whereby they can determine their migratory direction,
they are still reliant upon visual cues for their guidance. Although the earth's
magnetic field may provide important information concerning the compass
direction for migratory orientation, the available experimental data suggests
that this is used to calibrate visually based compasses (stars and position of the
setting sun) rather than referred to directly, either before or during the
migratory flight (Wiltschko and Wiltschko 1988b). However, night migrating
birds may not be visually well equipped to travel at night. Indeed, while the
attraction of night migrating birds to man-illuminated structures needs
explanation, the phenomenon would seem to suggest that these birds can be
subject to considerable perceptual confusion at night and that although
celestial cues may be available these may be overridden by cues from the
ground.

Fourthly, it is important to note that night migrating birds are flying at
considerable altitude, well away from obstacles and are thus not being called
upon to conduct the kinds of fine, visual, spatial discriminations which would
seem to be required when foraging by day. The detection of topographical
cues, such as river systems, coasts and mountain valleys would not require the
resolution of fine spatial detail.

NIGHT SINGING

One of the most enjoyable experiences of European ornithology is to listen to
the song of a Nightingale *Luscinia megarhynchos* in the dead of night. Even
though these birds can be heard singing during the day the quality of their
song is often regarded as even finer at night when they alone hold the stage and
other day-time sounds are gone. That Nightingales sing better at night is not
just a product of human perception abetted by romantic notions. Studies have
shown that the night-time song is more variable, composed of longer phrases
and more sustained; it may also contain phrases rarely heard during the day-
time (Hultsch and Todt 1981). Such night singing is another example of an
occasional nocturnal activity in an otherwise diurnal bird. Although night
singing as an activity would seem to bear little resemblance to night migration,
singing at night exhibits certain features which suggest something of the
limitations that night-time imposes on the activities of birds.

Although, within Europe, the Nightingale and its close relative the Thrush
Nightingale *L. luscinia* are justly renowned for their nocturnal singing, it is a
phenomenon which is not exclusively theirs. Among the Passeriforme birds
there are a number of species in which the males may sing during the night as
part of their breeding behaviour. However, nocturnal singing does not appear
to perform the same functions as singing during the day; it may in fact seem to
fulfil a specific and narrow function associated only with pair formation.

The nocturnal song of the Nightingale *Lusinia megarhynchos* is usually delivered from thick cover and appears to be solely concerned with mate attraction.

The best studied species in this respect, not surprisingly, is the Nightingale, with less detailed studies on the Thrush Nightingale. Work in this area has recently been reviewed (Cramp 1988, pp. 621 and 632). However, it is not certain how functional differences between night- and day-time singing might apply generally to the nocturnal and diurnal singing of other passerine species.

In the Nightingale, night singing seems to act as a long-distance advertisement to females, who tend to arrive in the breeding areas later than the males. Day-time song is mainly concerned with interactions between rival males, usually involving disputes over territorial boundaries. Early arriving males seem to sing exclusively during the day whilst establishing their territory. This accomplished, they begin to sing at night. Thus, nocturnal song is only associated with pair formation, and birds which pair successfully only sing at night for a brief period during the breeding season. One study in Austria indicated that males which paired successfully sang at night for only approximately 15 days whereas unpaired males sang far longer (Grull 1981). On wintering grounds in Africa nocturnal song is rare although day-time singing is commonly used in the defence of territories throughout the winter.

One further important difference between nocturnal and day-time song lies in the use of song perches. By day the male Nightingale may sing from several song perches, between which it alternates frequently. Nocturnal song, on the other hand, is given mainly from one perch which is typically used throughout the night for several nights in succession.

In the Thrush Nightingale it is believed that the functional roles of nocturnal and diurnal song are perhaps the opposite of those in the Nightingale. Thus, courtship song is usually delivered by day from a variety of low perches, whereas territorial song is typically delivered at night. However, as in the night-time singing of the Nightingale, a higher fixed song-perch is typically used. Although the birds sing frequently in their East African winter quarters, nocturnal song is absent.

Thus in both species nocturnal song occurs for only a short period of the annual cycle when it is associated with specific function. More importantly, the birds sing at night from fixed perches and are not generally mobile; there is no evidence that song bouts are interspersed with foraging or other activities. In short, the birds stay in one fixed position and do little other than sing.

Although systematic species lists are not available there are other Passeriforme species which are known to sing at night, though usually less reliably and less continuously than the two nightingale species. In none of these other species is singing exclusively a night-time activity. Singing occurs in intermittent bouts throughout the 24 hours for a relatively short period whilst establishing a breeding territory and attracting a mate. As in the Nightingales, individual birds may sing at night for only one or two weeks each year.

Among the Old World warblers (Sylviidae), species from the Acrocephalus genus, such as the Marsh and Sedge Warblers *A. palustris, A. schoenobaenus* or Locustella species, such as the Grasshopper Warbler *L. naevia* occasionally sing at night while among the thrushes (Turdidae), as well as the nightingales, the Wood Thrush *Hylocichla mustelina* of North America also falls into such a category. Another North American night-singer is the Northern Mockingbird *Mimus polyglottos* (Mimidae). Certain populations of Emberizidae, such as the Reed Bunting *Emberiza schoeniclus* may also sing after sunset, especially in northern populations in Sweden (Astrom 1974), although at such latitudes night-time light levels may remain relatively high throughout the spring and summer period when these birds are breeding (Chapter 2). The Wood Lark *Lullula arborea* is also noted for occasional night singing, although apparently only under moonlight (Cramp, 1988, p. 179).

Vocal behaviour at night is not restricted to the passerines. An example of a group of non-passerine birds which may be heard calling at night is the Rallidae (Gruiformes). Among them are such species as the Corncrake *Crex crex*, Spotted Crake *Porzana porzana* and Water Rail *Rallus aquaticus*, which in migratory populations may call intensively, day and night, after arriving on the breeding grounds. In the Corncrake the characteristic "crex-crex" call of the male probably has more than one function. It may serve as an advertisement call announcing the establishment of a territory, as a vocal challenge to intruders, and also serve to attract females. The bird may call for hours at a time but the frequency of song bouts declines as the breeding season progresses and the birds do not call whilst on their wintering grounds in East Africa (Cramp and Simmons 1980, p. 576). Another non-passerine noted for occasional night calling is the Cuckoo *Cuculus canorus*, although the function and conditions under which this occurs are not understood (Wyllie 1981).

As with nocturnal migration the actual behaviour performed by night-singers and night-callers would not seem to require the birds to make subtle visual discriminations. Typically, singing birds remain within a small area, frequently much smaller than the defended breeding territory, often on a single song-post. Although the vegetation in these sites may be dense

The scientific name of the Corncrake *Crex crex* reflects its monotonous call which may be heard throughout the night from damp meadows in many parts of Europe and central Asia.

(Phragmites reed for Acrocephalus warblers; scrub and low bushes for Nightingales and Grasshopper Warbler, and long grasses or reeds for Rails), it is generally rather uniform in structure. If the birds do move about whilst singing they appear to do so slowly, without recourse to flight. Thus singing at night does not appear to require a bird to make the same rapid and subtle visual discriminations which it may need to do when feeding or when involved in other aspects of its breeding behaviour during daylight.

NIGHT ATTENTANCE AT NESTS

The island of Skomer, off the coast of Pembrokeshire in Wales, during the summer months, holds a breeding population of nearly 100,000 pairs of Manx Shearwaters *Puffinus puffinus*. However, a day-time visitor to the island may see very little, if anything, of those birds on land. It is not until night-time that they come ashore to enter burrows and feed young or change incubation shifts. Manx Shearwaters on Skomer are one species among a group of marine birds which are noted for their arrival and departure from breeding colonies during the hours of darkness.

Most prominent among such birds are the smaller species in the orders Procellariiformes (the family Procellariidae, which includes the shearwaters and petrels; and the family Hydrobatidae, the storm-petrels), some species of penguin (Sphenisciformes), for example the Little Blue Penguin *Eudyptula minor* whose nightly return to a breeding colony near Melbourne, Australia, has become a tourist attraction, and the Swallow-tailed Gull *Creagrus furcatus* of the Galapagos.

In these colonial breeding species eggs are laid usually in an underground burrow or within a natural rock crevice. Entering and leaving these sites only at night is regarded as a strategy to avoid aerial predators. These sea-birds are particularly vulnerable to such attacks, especially the smaller species of the flightless penguins and the Procellariiformes whose anatomical adaptations to an aquatic life style have made their movements on land slow and cumbersome. Exactly which species among the Procellariiformes visit nests exclusively at night is not clear since, due to the remoteness of their colonies, the biology of many species is imperfectly understood (Bourne 1985).

Some of these birds are reported as coming ashore to the nest site only under the darkest conditions, e.g. Manx Shearwaters are less likely to visit the nest site if a moon is present (Harris 1966), although other species are apparently unaffected by the presence of the moon, e.g. the Storm Petrel *Hydrobates pelagicus* (Scott 1970) quoted in Cramp and Simmons 1977, p. 166.

The Swallow-tailed Gull is particularly interesting since it is the only gull species which is regarded as nocturnally active. Its breeding biology, behaviour and taxonomic status have been the subject of considerable study (Hailman 1964; Nelson 1968; Snow and Snow 1968; Harris 1970; Snow and Nelson 1984) but its nocturnal activities have not received detailed attention.

Manx Shearwaters *Puffinus puffinus* come ashore to nesting colonies almost exclusively under the cover of darkness.

The nightly return of Little Blue Penguins *Eudyptula minor* to their nest burrows has become a floodlit tourist attraction at one of their south Australian colonies.

Because of specializations in both morphology and behaviour (including its partially nocturnal habits) the relationship of the Swallow-tailed Gull to other Laridae has been uncertain. Some authors have regarded it as the sole member of the genus *Creagrus*, endemic to the Galapagos archipelago (Snow and Nelson 1984). Other authors have placed the Swallow-tailed Gull in the genus *Larus* and small numbers also breed outside the Galapagos on islands close to the mainland of South America (Harrison 1983).

Whatever its status there seems to be good agreement that the Swallow-tailed gull is unique among the Laridae in its nocturnal habits. While Hailman (1964) regarded the gull as entirely nocturnal the bird would not appear to fall within the definition of a nocturnal species presented in Chapter 3. For example, Nelson (1968) and Snow and Snow (1968) describe a wide range of display behaviours associated with breeding which were observed during day-time. Also, outside the breeding season the birds may wander over a wide area of the Pacific Ocean in the region of the Humboldt Current where they can be observed feeding by day. For this reason the behaviour of this gull is best regarded as occasionally nocturnal. However, there is some evidence that at least during the breeding season adults do forage on fish and squid at night (Harris 1970).

The nocturnal behaviour of the Swallow-tailed Gull parallels that of the Manx Shearwater in that it occurs principally during the breeding season when adults must return to land to provide food for the chick. Hailman (1964) and Snow and Nelson (1984) have pointed out that the home-coming birds are particularly vulnerable during the day-time to chasing by frigatebirds (*Fregata magnificens* and *F. minor*) which force the gulls to disgorge food. Visiting nests under the cover of darkness reduces the probability of such attacks. Furthermore chicks are themselves vulnerable to attack by frigatebirds and

Swallow-tailed Gulls *Creagrus furcatus* of the Galapagos are the only gull species which are regularly active at night.

the preferred nest sites, unlike those of most gull species, are on cliff ledges and in recesses. Where the birds nest in more open situations the site is usually near or under a boulder or bush thus making approach by the highly aerial frigate birds difficult. However, these nests are still vulnerable to attack by other species such as the Galapagos subspecies of the Short-eared Owl *Asio flammeus galapagoensis* (Nelson 1968) and the Galapagos Hawk *Buteo galapagoensis* (Hailman 1964). One difference between this gull and the shearwaters is that in the gull it is the chicks which are particularly vulnerable to predation either directly upon them or upon their food supply. The adults seem safer since they can feed out at sea away from the attentions of the frigate birds. In the shearwaters the adults as well as the chicks may be injured or killed by predators when on land.

Swallow-tailed Gulls show a number of morphological features which may be correlated with a more pelagic life-style and possibly with nocturnal activity. The wings are long and the legs short giving the bird somewhat tern-like proportions which presumably assist in long-distance flight over the ocean (the birds are often recorded several hundred kilometres from land). The birds are thought to subsist on a diet of squid and flying fish which Snow and Nelson (1984) speculate may be caught by plunge diving. Unfortunately there seem to be no observations of feeding at night-time in these birds.

As regards the nocturnal habit, Hailman (1964) noted that the eyes were unusually large for a gull and that they may contain a tapetum (see Chapter 7). He also drew attention to the markings on the adult's bill which could indicate its orientation in the dark and hence serve to guide the begging chick. The bill is blackish with a pale grey tip and there is a patch of white feathers at the base which contrast with the dark feathering of the rest of the head.

THE SENSORY BASES OF NEST LOCATION

Nothing seems to be known of how the Swallow-tailed Gull is able to locate its nest site at night. However, the sensory capacities associated with the Procellariiform species' ability to locate their nest sites at night has received considerable investigation. The studies, however, have been without definitive conclusion since no one experiment yet conducted has been able to control all of the possible cues (olfaction, audition and vision) in a systematic fashion; also, only a relative handful of species have been studied. That the Procellariiformes have a well developed sense of smell has been well established (Bang 1966; Bang and Wenzel 1985) and the possibility that they could use this sense to locate their individual nest burrow has been considered a number of times.

That some of the species are able to locate food sources in mid-ocean by smell alone has been demonstrated [Grubb (1972); Hutchison and Wenzel (1980); Hutchison *et al* (1984), cited in Bang and Wenzel (1985)]. Also, many of these species have a characteristic odour, and Grubb (1974) showed that Leach's Storm Petrels *Oceanodroma leucorrhoa* could discriminate between their own nest material and similar litter gathered from a forest floor, apparently on the basis of olfaction. These and other experiments and observations led Grubb (1974) to conclude that Leach's Storm Petrel used olfaction not only to locate the island colony but also to locate individual nest burrows in darkness. This is the only species for which such a conclusion has been drawn, but the birds of the colony tested by Grubb differed in one important respect to many other Leach's Storm Petrels (and many other species of Procellariiformes) in that their nest burrows were among tree roots under a closed canopy of conifer forest. The nesting colonies of many Procellariiform species tend to be in open situations well away from the shade cast by a tree canopy which, as discussed in Chapter 2, can reduce ambient light levels by up to 100-fold.

Studies of burrow location in other species of Procellariiformes have generally led to the conclusion that visual cues are employed, though no one study has definitively demonstrated that visual cues alone are used by all birds to find their individual burrows. Among the species for which this conclusion has been reached are: Manx Shearwater (Brooke 1978a; James 1986); Wedge-tailed Shearwater *Puffinus pacificus* (Shallenberger 1975); Little Shearwater *P. assimilis*, Pale-footed Shearwater *P. carneipes* and Short-tailed Shearwater *P. tenuirostris* (Warham 1955, 1958, 1960).

A study of Cory's Shearwater *Calonectris diomedea* by Wink *et al* (1980) concluded, however, that in this bird there was evidence that individuals employed echolocation to assist in locating the nesting burrow as they flew in from the sea. This echolocation apparently used vocalisations whose frequencies were within the normal range of human hearing. Indeed Ranft and Slater (1987) failed to find any evidence that Storm Petrels produced sounds in the range of ultrasonic frequencies which are employed by various species of bats in their echolocatory guided avoidance of obstacles. However, failure to detect such high frequency sounds does not rule out the possibility that these birds could be using echolocation as an aid to successful homing to the nest site. As discussed in Chapter 6, birds which are known to be able to guide their flight by echolocation do not employ ultrasonic frequencies.

A further way in which hearing could be used to assist in the location of the nesting burrow is through the recognition of the call of the mate inside the burrow. That such a cue is available to Manx Shearwaters was demonstrated by Brooke (1978b) who showed that individuals could recognise each other's calls. However, this cue is not essential for burrow location since birds can find the correct burrow when it is occupied by only a silent chick.

Night attendance at the nest in Procellariiform birds has been regarded as a "remarkable ability" (Ranft and Slater 1987), and indeed this is how it may seem when experienced on a remote island colony, when literally thousands of birds may come ashore almost simultaneously to visit their nesting burrows under the cover of darkness. However, when viewed within a wider perspective this behaviour is perhaps no more remarkable than the ability of any bird to home to its nest site in a dense colony or over a long distance.

These Procellariiform birds have available to them a range of cues based upon vision, audition and olfaction which could be used in an heirarchical or redundant manner, in much the same way that the various sensory cues complement each other in long-distance navigation (see above). For example, it may be that Leach's Storm Petrels on the wooded island where Grubb (1974) studied them, use olfaction to locate their nesting burrows because of the particular circumstances there (heavy shading produced by a tree canopy), but at other sites vision is the primary cue employed, as was demonstrated by Brooke (1978a) in the Manx Shearwater. However, even Manx Shearwaters are likely to bump into each other and into obstacles as they fly in towards their nesting burrows (Brooke, personal communication).

Certainly, studies of homing in Pigeons have shown convincingly the futility of attempting to find a single cue upon which birds will rely in order to regain their nest site. These birds, too, have available to them a range of visual, olfactory and auditory cues which may be employed in order to home successfully to a nest site. Which cues are used or revealed depends upon the circumstances or experimental manipulatons conducted by the investigator [see, for example, the summaries and reviews of Keeton (1979a,b) and Baker (1984)]. It would seem that the ability to home to a nest site is too vital a requirement for it to be entrusted to a single cue; redundancy of cues is the order of the day (or night). It would seem likely that a similar situation would apply to all birds which had to regularly home to their nest site from a considerable distance, as is the case in the Procellariiformes.

Two further general points follow from this discussion. First, if the difference between the primary reliance upon olfactory and visual cues for nest location, in Leach's Storm Petrel and Manx Shearwater respectively, does depend upon the presence of a tree canopy at the former species' colony, then it may be suggested that the nocturnal vision of these birds is only sufficient to cope with the visual problems posed at night in open habitats [Grubb (1974) actually describes Leach's Storm Petrels bumping into large obstacles on their way to their nest burrows]. Entering beneath the shade of a tree canopy is sufficient to reduce visual perception to a point where that cue becomes unreliable or unavailable. Although olfaction is perhaps, in general, a less reliable or less rapid cue than vision for the exact spatial location of a nest site, it nevertheless can function adequately to locate the nest site when visual guidance is no longer possible. This analysis parallels that applied to the situation in homing pigeons, where only certain populations have been found to utilise olfactory cues in homing (Papi 1982). There are other species of Procellariiformes, especially among the group known as the gadfly petrels (*Bulweria* and *Pterodroma* spp.), which apparently visit nest sites by night, often in forested mountain slopes far inland (Bourne 1985, p. 454). Grubb (1974) lists ten species of shearwaters and petrels which apparently locate their nests at night under a tree canopy. Clearly when and how these species find their nest sites could be of particular interest to this discussion.

Secondly, it should be noted that as in the case of nocturnal migration all flight takes place in open, structurally simple, habitats, where natural obstacles are few and presumably fine spatial judgements are not required. Even in the case of the petrels nesting beneath trees the birds do not seem to fly far, if at all, beneath the tree cover. They tend to crash through the tree canopy, land and then walk to their nest burrows (M. Brooke, personal communication).

OCCASIONAL NIGHT FORAGING

Migrating, singing or returning to a nest site during night-time are relatively conspicuous activities which have been recorded many times by a variety of means. Discovering which species are involved and the sensory basis of their behaviours is problematic. They are certainly not comprehensively understood, but at least investigators have been able, without too much difficulty, to assemble systematic observations and devise testable hypotheses concerning the possible sensory bases of such occasional night-time activities. Feeding by night, however, presents more difficult problems and more uncertainty surrounds the understanding of this occasional nocturnal behaviour.

While it may be possible to show that a bird is present in a possible feeding area at night, it is much more difficult to be certain what the bird is doing and how it is doing it. From the sensory point of view there are two main problems: how to account for the bird's mobility at night and how to account for its ability to locate food items. The former question may in large part be answered by the fact that occasional night feeding in all species takes place in essentially open habitats, devoid of obstacles, which are usually associated with large bodies of water. Although the birds are not flying in a completely open air space, like the nocturnal migrants, they are nevertheless not required to make fine spatial discriminations during flight or when landing. Also, the feeding sites are likely to be traditional and well known to the individual birds which feed there both night and day. To understand how these birds actually locate their food during night-time, reference has to be made to sensory capacities often overlooked in birds; olfaction, taste and touch sensitivity and, possibly, hearing.

PELAGIC SEABIRDS

Analysis of stomach content may be the only way of knowing that a bird has been feeding by night, and it is mainly on the strength of such data that night-feeding is believed to occur in some of the pelagic seabirds (Imber 1973; Clarke *et al* 1981; Croxall *et al* 1988; Harris 1970). The actual whereabouts and feeding activities of such species even during daylight, let alone at night-time, is practically impossible to determine with certainty, although some data is available from radio tracking studies of individual birds (Prince and Francis 1984). However, observers have pieced together information which suggests that at least some of the Procellariiform species and the Swallow-tailed Gull do indeed feed occasionally, perhaps regularly, at night, at least during the breeding season. However, for many species there is little or no specific information on the diet or when food is obtained.

Examination of the data collated by Cramp and Simmons (1977) on the Procellariiform species which occur in the Western Palearctic shows that for many species little is known about the timing and techniques used in feeding. It is, however, clear that night attendance at the nesting colonies does not necessarily imply that individuals of these species also feed at night. Thus in describing the feeding of the Manx Shearwater *Puffinus puffinus* it is stated that the birds feed "by day, by pursuit-plunging, pursuit-diving and by surface-seizing" (Cramp and Simmons 1977, p. 148). This does not of course imply that the birds never feed at night, but the fact that night-time feeding is not mentioned suggests that if it does occur it is relatively infrequent. The same pattern of night attendance at the nesting colonies but feeding during day-time is apparently found in the Sooty Shearwater *P. griseus*, Great Shearwater *P. gravis* and Storm Petrel *Hydrobates pelagicus*. Nocturnal feeding, however, is reported to occur in Bulwer's Petrel *Bulweria bulwerii*,

Storm Petrels *Hydrobates pelagicus* may be encountered feeding by both night and day over wide areas of the Atlantic Ocean and Mediterranean Sea.

Cory's Shearwater *Calonectris diomedea* and possibly in Leach's Storm-petrel *Oceanodroma leucorrhoa*.

The feeding technique used does not seem to be correlated with occasional nocturnal feeding in these species. The Procellariiformes feed on invertebrates, fish or plankton, taken either from the water surface or underwater using shallow surface or plunge dives; the birds stay underwater for only a few seconds. Cory's, Manx, Sooty and Great Shearwaters are reported to use all of these techniques but only one of the species, Cory's Shearwater, is regarded as an occasional night-feeder.

On the other hand, the Storm Petrel and Leach's Storm-petrel feed by snatching items from the sea surface whilst in flight, hovering or pattering over the surface (both species rarely land on water to feed and apparently never dive), yet the Storm Petrel is regarded as a day-time feeder [Cramp and Simmons (1977), p. 165, but see the results of Grubb (1972) discussed below] while there is evidence that the Leach's Storm-petrel may feed by both day and night. It is of course possible that these birds may use different feeding techniques during night- and day-time, or that they feed at night only under the higher levels of natural nocturnal illumination, but there is no information available on either of these points.

Analysis of the diet (stomach contents) of various species of the larger Procellariiformes (family Diomedeidae, albatrosses) and the Swallow-tailed

Gull also indicates that some of these species may feed at night. The evidence is based on the fact that the diet of these birds includes animals, particularly squid, which exhibit a vertical migration during the daily cycle, which brings them within the surface feeding range of the birds only during the night. These studies have shown that the Wandering, Black-browed, Grey-headed and Yellow-nosed Albatrosses (*Diomedea exulans, D. melanophris, D. chrysostoma, D. chlororhynchos*) (Clarke *et al* 1981, Clarke and Prince 1981; Brooke and Klages 1986) and the Great-winged Petrel *Pterodroma macroptera* (Imber 1973), all appear to feed during the night, though it seems unlikely that they do so exclusively. It is of interest to note that although these albatross species may feed at night their nests are typically in open situations, and they do not rely on the cover of darkness when visiting the site in the way that the smaller species of petrel do. This is presumably because their large size is sufficient to protect them from aerial predators.

Sensory cues to prey location

Grubb (1972) showed that during day-light, vision and olfaction may both be used to detect possible food sources in some species of Procellariiformes. He placed sponges soaked either in cod liver oil or seawater just above the sea surface and observed that birds from four species (Great and Sooty Shearwaters, Wilson's and Leach's Storm-petrels) were attracted to the baits. All of the species, except the Sooty Shearwater, showed a preference for the pungent oil-soaked sponge. Five species from two other orders of seabirds, Pelecaniformes and Charadriiformes, were present in the area of the tests but none of them responded to either of the sponges; Gannet *Sula bassana*, Great Black-backed Gull *Larus marinus*, Herring Gull *L. argentatus*, Arctic Tern *Sterna paradisaea* and Puffin *Fratercula artica*.

When the same test was conducted at night, however, only the Storm Petrels were attracted to either of the baits and they showed a strong preference for the oil-soaked sponge suggesting that they were guided primarily by olfactory cues. Thus it would seem that the two species of Storm Petrels may be actively foraging at night and that olfaction could play an important part in this. In the two species of shearwater, however, foraging at night may not occur, or at least it would not seem to be guided by olfactory cues. These conclusions are in good agreement with possible differences in the use of olfactory and visual cues in the location of nests at night between the Manx Shearwater and Leach's Storm Petrel discussed above. There is no data on how Swallow-tailed Gulls can locate squid at night.

The picture presented here is somewhat confused as regards which species of pelagic seabirds do what at night. However, a few clear points do emerge. First, there seems no doubt that some of these seabirds do feed at night and that olfaction can play an important role in locating food items. It would even seem that olfaction could aid some species to locate particularly pungent individual food items, as well as bringing them to a general area in which food is available. Secondly, the degree to which different species employ olfaction is not clear but this is hardly surprising given the great difficulty of studying these oceanic birds. Thirdly, as was noted above when discussing night attendance at the nest in these species, all flight takes place in open situations well away from any obstacles. This final point, when linked with the use of olfactory cues, would suggest that high visibility is not essential for these birds when foraging, either by night or day.

WADERS AND WATERFOWL

Two other groups of birds are particularly noted for their occasional nocturnal foraging, the waders or shorebirds (Charadriiformes) and the waterfowl or wildfowl (Anseriformes). As is the case with nocturnal migration, although there are many casual observations of night-feeding there are insufficient definitive studies to produce systematic tables of its occurrence across species. Also, very little is known of the particular circumstances, especially light levels, under which certain species chose to feed at night. Possible differences in the way that birds may forage by night and day, and the efficiency of nocturnal versus diurnal feeding, are known for just one or two species.

Nocturnal feeding in waders and waterfowl occurs especially outside the breeding season when these birds are most gregarious and present in temperate areas on lakes, sea coasts and estuaries. When birds are present on coasts and estuaries the times available for roosting and feeding may be dictated primarily by the tidal cycle and secondarily by weather conditions (Evans 1976; Hale 1980). Where birds are feeding at sites affected by the tidal cycle, individuals may simply continue foraging on available areas through twilight into the night-time light levels, and stop only when forced to do so by the incoming tide. However, it should not be assumed that all species do this no matter what the circumstances. Indeed there is evidence that the feeding cycle may be a diurnal–nocturnal one modified by the tides as, for example, in the Oystercatcher (Heppleston 1971), and that nocturnal feeding occurs only under certain conditions. This suggests that feeding at night, at least among some species of waders and waterfowl, may be a less preferred strategy than feeding by day.

The suggestion is supported by observations that night feeding is more likely to occur in mid winter than at other times of the year; for example, in Oystercatchers *Haematopus ostralegus*, Bar-tailed Godwits *Limosa lapponica*, Grey Plovers *Pluvialis squatarola* and Redshank *Tringa totanus* (Goss-Custard 1969; Heppleston 1971; Pienkowski 1982; Cramp and Simmons 1983, pp. 21, 477 and 532). It is assumed that this is a response to generally more difficult feeding conditions at all times of the day, and the fact that winter days are relatively short, rather than a particular preference for night-time feeding. However, Dugan (1981) has suggested that Grey Plovers may also feed at night because night-time brings an increase in the availability and activity of prey.

In these same species and in many others, for example, the Avocet *Recurvirostra avosetta* and Lapwing *Vanellus vanellus*, it has also been recorded that night-feeding is more likely to occur on moonlit nights (Spencer

Bar-tailed Godwit *Limosa lapponica*, one of many long-billed shorebirds which can probably locate prey exclusively by using touch sensitive receptors in the bill tip.

1953; Goss-Custard, personal communication). In the Oystercatcher it has been shown that feeding rates at night are about half those recorded on the same foraging area the following day (Goss-Custard and Durell 1987). However, in the Grey Plover feeding rates were found to be significantly reduced on moonless nights, but not on moonlit nights, compared with feeding rates during the day (Pienkowski 1982). It should not be assumed that night feeding in waders occurs exclusively in birds wintering in temperate latitudes, where weather conditions may be harsh and hence food requirements temporarily high. Robert and McNeil (1989) and Robert *et al* (1989) have shown that wading birds wintering in the tropics may also feed at night, but they propose that even at these latitudes such nocturnal feeding is associated with particularly high food requirements. For example, at the time of pre-migratory fattening, or refuelling at a stop-over site, or after a long non-stop flight, rather than as a general strategy adopted by the birds during their non-breeding period.

Nocturnal feeding may occur on a seasonal basis even in wading species whose activities are not governed by the tides. Radio-tracking of individual birds (Hirons 1981, cited in Cramp and Simmons 1983) has shown that in the case of the Eurasian Woodcock *Scolopax rusticola* nocturnal feeding occurs mainly during the winter when birds forage for soil invertebrates in pastures outside of the woods where they roost during the day. During the breeding season the birds tend to forage only by day on the woodland floor. It seems likely that a similar foraging pattern also occurs in the American Woodcock *S. minor* (Sheldon 1967; Stribling and Doerr 1985). This seasonal difference in the Woodcock's nocturnal feeding is in agreement with the fact that when feeding by night, both waterfowl and waders prefer open habitats which are largely devoid of obstacles.

Among the waders and shorebirds night-feeding occurs principally among the longer-billed species which forage by probing, rather than the short-billed species which feed by picking individual items from the substrate or snatching insects from the air. Thus species like the Little Ringed Plover *Charadrius dubius* and Kentish Plover *C. alexandrinus* appear not to have been recorded feeding by night (Cramp and Simmons 1983, pp. 118 and 156) and only occasionally do species such as the Golden and Grey Plovers *Pluvialis apricaria* and *P. squatarola* and Lapwings, which usually take items from the surface, feed at night (Spencer 1953; Pienkowski 1982).

The following brief survey of the incidence of occasional nocturnal feeding among the waterfowl (Anseriformes) will serve to point up how some of the above-mentioned factors are associated. The data used here are based upon Cramp and Simmons (1977) under the headings "Food" and "Roosting" for each of forty species in the Anseriformes. Although Cramp and Simmons probably present the most comprehensive review available, data on the occurrence of night-feeding are fragmentary. For two species, Mandarin *Aix galericulata,* and Ferruginous Duck *Aythya nyroca,* data under the two headings are contradictory.

Anserini (swans and geese). As a rule all species roost at night and feed by day, with occasional nigh-feeding rarely recorded. There is, however, one exception to this, the Brent Goose *Branta bernicla.* This is the only species in the tribe habitually to feed in tidal areas during the non-breeding season and the only one which is recorded as regularly feeding at night according to the tidal cycle, but even then birds aparently do so only on moonlit nights.

Tadornini (sheldgeese, shelducks). The Egyptian Goose *Alopochen aegyptiaca,* like many of the geese and swans from the Anserini is a freshwater species which feeds by grazing. However, unlike them it apparently feeds almost exclusively at night. On the other hand the Shelduck *Tadorna tadorna* is a coastal species which feeds at night according to the tides.

Anatini (dabbling ducks). All of the species in this group which frequent estuaries are known to feed at night depending upon the tides, for example, Wigeon *Anas penelope,* Teal *A. crecca,* Mallard *A. platyrhynchos* and Pintail *A. acuta.* Birds of these species which feed in places not affected by tides, do so primarily by day, but even these birds may feed regularly at night if they are persecuted or disturbed by man or natural predators.

Aythini (pochards). All of these species of diving duck will feed at night, even in non-tidal situations and when not persecuted or disturbed by day, and when a moon is not present. Their food is primarily vegetable matter and small invertebrates, so it seems highly unlikely that food becomes more easily available at night-time. The Pochard *Aythya ferina* and Ferruginous Duck *A. nyroca* seem to feed mainly at night whatever the conditions, while the Tufted Duck *A. fuligula* and Red Crested Pochard *Netta rufina* are more likely to be recorded feeding during the day.

Touch and taste sensitivity in the bill enable Mallards *Anas platyrynchos* to locate and identify food items buried beneath soft surfaces.

Somateriini (eiders). Unlike the previous group of diving ducks these birds usually feed by day except where tidal cycles, as in the case of Eiders *Somateria mollissima* in northern Britain, influence feeding.

Mergini (scoters and sawbills). All of these diving birds, which feed on molluscs and fish, are regarded as day-time feeders which always roost throughout the night. Although they often feed in coastal situations there is no evidence that they are influenced by tides.

It can be clearly seen from the above brief survey that feeding at night for most species in the Anseriformes is best regarded (as in the case of nocturnal migration) as a flexible response to a local situation (tides, day-time disturbance, winter food shortage) rather than a preferred strategy for which the animal is well prepared. It may be best to consider that in these birds, and in the wader species, the feeding technique and the habitats in which they feed, are such as to *permit* feeding at night should that become necessary due to energetic considerations, rather than that the birds have a preference for night feeding. The sensory basis of the feeding techniques which permit this occasional night feeding are the subject of the next section. There is, however, one particular problem which cannot be explained or be so easily accounted for, that is, the apparent preference for feeding by night reported in the Pochard and in the Egyptian Goose. Two quite different feeding techniques and diets are involved in these two birds so it seems unlikely that there is a common factor operating. More data on the feeding of these two birds would clearly be of value.

Sensory cues for food location

It seems likely that the detection and capture of food items in nocturnally feeding waterfowl and waders can rely exclusively upon tactile and taste cues. Auditory cues may play an additional role in at least some species of waders. Where visual cues are used for feeding at night they are usually employed only when night light levels are high, particularly at the time around full moon.

Curlews *Numenius arquata*, like many other long-billed shorebirds, may feed by both night and day when visiting estuaries.

There is also evidence that some species will change their foraging technique as a function of light level.

Both waders and waterfowl species are noted for the variety of feeding techniques which any one species may use (see Cramp and Simmons 1977, 1983; Hale 1980). Field observations suggest that different techniques depend to various degrees on hearing, visual or tactile cues. This is exemplified well by observations of the Curlew *Numenius arquata* which uses three main feeding techniques. (1) Pecking, in which the bill merely touches the surface. (2) Jabbing, in which the bill is rapidly inserted and withdrawn but only up to about half its full depth. (3) Probing, which is a longer movement in which the bill is partly or fully inserted into the mud. Presumably the first technique involves visual and/or auditory guidance while the latter two must depend to varying extent upon tactile cues; indeed it has been recorded that the Curlew can locate and swallow whole, small bivalve molluscs without withdrawing its bill from the mud.

Many long-billed waders are able to seize prey with the bill tip whilst it is still buried in the mud or soil, using the so-called rhynchokinetic property of the bill which enables just its tips to be separated by the upward bending of only the distal portion of the upper mandible (Burton 1974; Pettigrew and Frost 1985). This is illustrated in the case of the Dunlin *Calidris alpina* in Figure 4.1.

There are many examples of stereotyped foraging techniques similar to those of the Curlew, which are conducted without the apparent benefit of visual cues. These include the Bar-tailed Godwit's so-called "stitching" action (a rapid series of probes which are shallow and close together) and the Oystercatcher's "sewing" movements (Cramp and Simmons 1983, pp. 21 and 476).

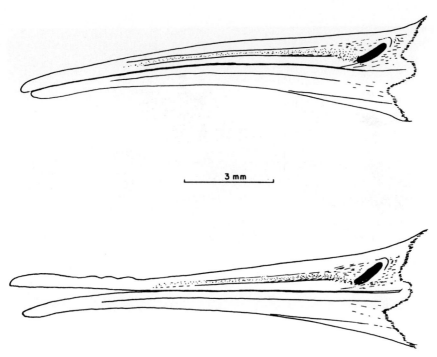

Figure 4.1 The bill of the Dunlin *Calidris alpina,* showing rhynchokinesis. In the upper diagram the upper mandible is curved slightly downwards along most of its length, and the soft flexible tips of the mandibles meet. In the lower diagram the terminal 5 mm of the upper mandible is bent slightly upwards thus separating only the tips of the mandibles. (From Pettigrew and Frost 1985.)

When foraging in these ways clues as to what lies buried in the mud are almost certainly provided by tactile receptors (principally Herbst Corpuscles) which are found concentrated at the bill tip (Bolze 1968; Gottschaldt and Lausmann 1974). These mechanical, or touch sensitive, receptors are not found exclusively in the bill tips of these birds but are found at many sites on the body surface of most bird species and even on the tongue tips of Woodpeckers. However, where these receptors are found concentrated at the tip (Figure 4.2), as opposed to along the edges of the bill, a primary role in the actual detection of prey is indicated (Gottschaldt 1985). It has been shown in the Dunlin that a relatively large area of the brain is devoted to the analysis of tactile information from the bill tip (Pettigrew and Frost 1985) and that this may be likened to a "tactile fovea" in the sense that it represents a focus of heightened sensitivity. Pettigrew and Frost suggest that such a tactile fovea may be common to all long-billed, probing, birds of the family Scolopacidae (sandpipers and their allies).

Touch sensitivity in the bill has been shown capable of controlling apparently more complex foraging tasks than those conducted by waders.

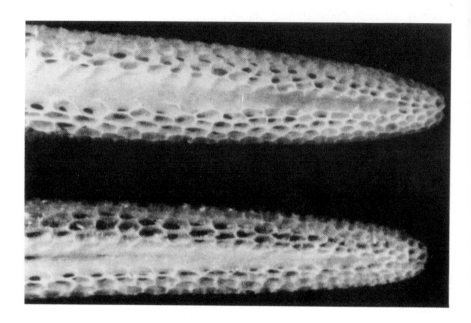

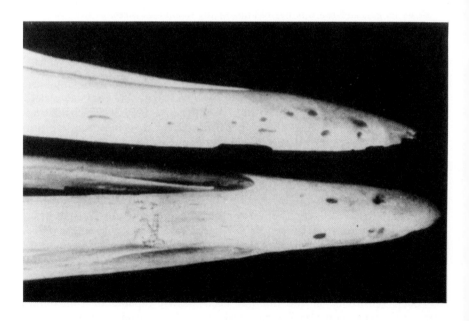

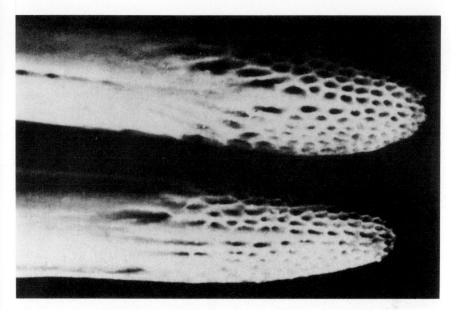

Figure 4.2 The tips of the bony parts of the upper and lower mandibles in three species of shorebirds (Charadriiformes), showing the small pits in which the touch sensitive receptors are located on the *outer surface* of the beak beneath the covering of the rhamphotheca, which in these birds is soft and leathery. The three species illustrated show marked differences in the number of pits. Opposite top, Dunlin *Calidris alpina*; bottom, Lapwing *Vanellus vanellus*; above, Snipe *Gallinago gallinago*. (From Bolze 1968.)

Interesting examples of this are found in two groups of unrelated species, the storks (Ciconiidae; Ciciniiformes) and the skimmers (Rynchopidae; Charadriiformes) and brief discussion of these will indicate the kinds of sophisticated and very rapid feeding movements which can be achieved by touch sensitivity in the bill alone.

The America Wood Stork *Mycteria americana* often searches for prey in turbid waters. Kahl and Peacock (1963) demonstrated that temporarily blindfolded birds could continue to catch live fish with no apparent decrease in efficiency or reaction time, the birds apparently detecting the fish when they touched the bill tip. The three species of skimmers are also thought to be able to detect their prey exclusively by tactile cues by day or night (Zusi 1962). The skimmers have a unique bill structure with which the prey can be both detected and caught (Erwin 1977). They fly horizontally with the bill open and the lower mandible (which is blade like and longer than the upper mandible) cutting through the water. The prey is detected as it touches the lower mandible and a very fast reflex action permits the prey to be snatched from below the water surface as the bird continues to fly. As in the case of the waders and waterfowl, skimmers are birds of open habitats, preferring wide lakes, rivers and coastal lagoons. Their foraging at twilight is discussed in more detail in Chapter 5.

Evidence that touch and taste sensitivity within the bill may be sufficient in itself to control foraging in waterfowl comes mainly from studies of the "bill tip organ". This organ was first described over a century ago in parrots (Goujon 1869) and then in other species, including the ducks. However, during the present century it was virtually forgotten until its "rediscovery" by Gottschaldt and Lausmann (1974). It has been described by Gottschaldt as an avian equivalent of the sinus hair system in the mole which guides that animal's activities underground in complete darkness. It has also been regarded as the most sophisticated sensory organ occurring in vertebrate skin.

The general structure of the bill tip organ appears to be similar in all avian species in which it is known to occur. In birds which use their bills primarily as a grain-pecking tool, such as pigeons and sparrows, the organ is absent. However, if the beak "is used in a more instrumental way to search for, catch, select or manipulate food – as, for instance, in parrot or waterfowl – a well-developed bill tip organ can be found" (Gottschaldt 1985, p. 446). Its structure in the domesticated (Greylag) goose is shown in Figure 4.3.

Figure 4.3   The bill-tip organ in the upper mandible of the Greylag Goose *Anser anser* at two magnifications. The positions of the receptors of the organ are indicated by openings in the hard horny rhamphotheca. These small tubules are arranged in rows around the edge of the bill *inside* the beak. (From Gottschaldt 1985.)

Basically the organ consists of a mass of touch sensitive receptors arranged in rows inside the rim of the upper and lower mandibles at their tips. There are usually more receptors in the lower than in the upper mandible. The touch receptors themselves are of four types, each responding to a different kind of stimulus and arranged in groups which are mechanically isolated from each other. This isolation, coupled with the very high density of receptors (up to 1,000 per mm^2), should enable the bird to make very fine tactile discriminations with its bill tip.

It is clear that touch sensitivity plays a crucial role in the feeding behaviour of waterfowl and wading species [see Gottschaldt (1985) for a thorough review of touch sensitivity, especially concerning its role in feeding, in these and other species of bird] and can permit them to feed without recourse to visual cues. However, while touch sensitivity is sufficient to enable birds to detect food items and manipulate them in their bill ready for swallowing, it is also essential for a bird to assess the palatability of food items before ingestion.

The importance of the sense of taste in birds is often overlooked but there is now a large body of data describing the physiological basis of taste reception and its importance in feeding behaviour. This has recently been reviewed by Gentle (1975), Kare and Rogers (1976) and Berkhoudt (1985). Although most work has concentrated on domesticated species, other studies have shown the importance of taste in the feeding behaviour of a wide range of bird species from a number of orders, including both the waterfowl and the waders.

Some studies have shown how sensitivity to tactile cues from the bill can work in conjunction with visual or taste cues to control the ingestion of food in birds (Zweers 1982a). However, there are marked differences between species concerning the importance of these cues. For example, it seems likely that the domestic fowl's preference for particular seeds is decided by tactile instead of taste cues (Engelmann 1957); in finches, however, it seems that visual properties and surface texture have precedence over all other qualities in the bird's selection of individual seeds (Morris 1955; Kear 1960). On the other hand, in waterfowl and wading birds there is evidence that tactile and taste cues take precedence over visual cues.

Perhaps the most clear demonstration of the primacy of touch and taste cues is provided by the experiments of Zweers and Wouterlood (1973). They devised a task in which Mallards were trained to forage for food items hidden in wet sand. When required to discriminate between real peas and plasticine imitations (same size, shape and colour) they detected, retrieved and swallowed all the hidden peas, leaving the fake peas buried in the sand. Thus not only were the birds able to recognise the peas from among the fakes but they did this when the beak tips and the food items were still under the sand. In addition Berkhoudt (1977, 1980) has shown how the areas inside a Mallard's mouth where touch and taste receptors are concentrated are close to the pathway followed by either strained or pecked food items.

Evidence that wading birds of the genus *Calidris* can use taste cues to determine the presence of prey in sand comes from equally intriguing experiments (Gerritsen *et al* 1983; Van Heezik *et al* 1983). These demonstrated that four species, Sanderling *C. alba*, Dunlin *C. alpina*, Purple

Sandpiper *C. maritima* and Red Knot *C. canutus,* can discriminate between samples of sand depending upon its taste.

In these experiments birds were first trained to forage for natural foods placed in jars of sand. Birds were then presented with pairs of jars, one containing "taste" the other "no-taste". "Taste" came from polychaete worms *Nereis diversicolor* or bivalve molluscs *Macoma balthica* which were introduced to washed sand but removed before the experiments. "No-taste" was simply the washed sand. It was found that all birds spent longer and probed more frequently at the jar containing "taste" than "no-taste". Gerritsen *et al* (1983) concluded that in all species taste sensitivity guided foraging but that its importance varied between species. It was suggested that the Dunlin is most dependent upon taste cues and the Knot the least.

Thus it seems that these species can determine which areas of apparently uniform substrate will be more profitable for foraging, using cues based solely on the strength of certain tastes within it. Gerritsen *et al* also conjecture that taste buds, as well as touch sensitive receptors, must occur at the tips of the beaks in the four species tested.

There is one further piece of evidence which suggests that waders and waterfowl can forage successfully in the absence of visual cues and this comes from studies of the visual fields of birds. The visual field is simply the space around the bird in which vision of some kind is possible. In birds which are known to be guided to food items by vision, such as the Pigeon *Columba livia* (Zweers 1982a) and European Starling *Sturnus vulgaris* (Beecher 1978), the bill is positioned well within the bird's frontal binocular field (Martin and Young 1983; Martinoya *et al* 1981; McFadden and Reymond 1985; Martin 1986c) and it is presumed that this enables visual guidance of the peck towards individual food items. In the Pigeon it has been shown that such pecks are under visual control to within a few millimetres of the target (Zeigler *et al* 1980; Zweers 1982b). However, in the case of the Mallard the bill tip falls on the very edge of its visual field (Martin 1986b), the eyes being positioned so as to give complete visual coverage of the celestial hemisphere above the bird's head (and vision directly behind the head) rather than to view the ground around and below the bill. Thus the bird is well equipped to detect predators (a mallard cannot be approached by a predator from any direction, except below, without being seen) but is only just able to see the position of its bill tip. This also suggests that visual guidance during feeding is not particularly important in a species which feeds primarily by grazing and straining food items from mud or water.

Finally, mention should be made of two studies which suggest that hearing could play a role in the foraging of at least some species of wader, notably the plovers of the family Charadriidae. All of these species feed occasionally at night but their foraging typically involves taking prey from the surface, or at least probing no deeper than a few centimetres into sand or mud. Compared with those of the deeply probing species of Scolopacidae, the bills of the Charadriidae are relatively short, between 20 and 30 mm long. Field observations suggest that their daylight foraging may be guided primarily by visual cues, but Fallet (1962) and Lange (1968) have shown experimentally that they may also use sounds. The cues would be the sounds of invertebrates as they move on or just below the surface of mud or soil. Certainly, sound cues to the presence of invertebrates in soil are potentially available to all bird species which feed on the ground, and it would seem unlikely that they would not be utilised if reliably present. It is unlikely that sounds could be the sole cues by which prey items are detected and Fallet has argued from experimental studies that Golden Plover *Pluvialis apricaria* and Lapwing *Vanellus vanellus* can use hearing to detect prey which is then searched for using the bill as a tactile probe. However, Peinkowski (1980, quoted in Cramp and Simmons 1983, p. 218) has argued from field observations that during daylight the use of sound cues in foraging is, at best, uncommon in these plovers. He suggests that when foraging on an estuary they are guided mainly by visual cues produced as the prey moves in the mud or disturbs shallow water.

Regardless of whether auditory cues are available it seems clear that tactile and taste cues must play an important role in the feeding behaviour of many, if not all, of these occasionally night foraging species of waders and waterfowl. It would seem that these cues can be sufficient in themselves to allow feeding in the absence of visual cues in many species of waders and waterfowl.

There is evidence that wader species can be flexible in their use of visual and tactile cues whilst foraging, according to the time of day, and that associated with this switch may be a change in the type of prey taken. Thus it seems likely that Oystercatchers use visual cues to select individual cockles and mussels from the surface during daylight, but switch at night to probing using stereotyped "sewing" movements in soft mud when, presumably, tactile cues will also be used to detect soft bodied prey (Hulscher 1976). Black-winged Stilts *Himantopus himantopus,* and Greater and Lesser Yellowlegs *Tringa melanoleuca* and *T. flavipes* were also found to switch from a day-time foraging technique, apparently guided mainly by visual cues, in which individual items were pecked at, to one guided mainly by tactile cues when feeding at night (Robert and McNeil 1989). Robert and McNeil suggest the birds may change their diet with the change in foraging technique, but were unable to provide evidence for this. It should be noted that all of these are long-billed birds which may thus have the "option" of changing their feeding technique and diet according to local light conditions. Such flexibility may not be open to short-billed waders such as the Plovers *Charadriidae* spp., which are more likely to be visually guided. This may explain why these short-billed birds are more likely to feed at night only during bright moonlight, though their feeding rate may be much reduced when they do so (Spencer 1953; Pienkowski 1982; Robert and McNeil 1989).

ARCTIC BIRDS

In Chapter 2 the annual light regime at high latitudes was briefly described. The great advantage to certain species of breeding at these high latitudes in the summer months would seem to be clear (Ogilvie 1976). There is practically no daily cycle of night and day, but long periods of continuous daylight during which there is an abundance of insect prey and rapid vegetation growth. At latitude 76°N the sun stays above the horizon for almost one third of the year from April 24th until August 19th.

The winter, however, is a different matter, and at this same latitude the sun stays continuously below the horizon for over 100 days (November 1st–February 10th). These long periods of night-time light levels are the very coldest time of the year and so it is surprising to find that any bird species remain at these latitudes throughout this period. Nevertheless, five species may spend the whole year resident at these very high northern latitudes. Also, where these five species occur elsewhere in their range they are not noted for their nocturnal activities. Thus they should be considered as occasionally nocturnal birds even though the "occasion" when they must be active at night-time light levels may last many weeks.

The Red (or Willow) Grouse *Lagopus lagopus* may be resident as far north as 70° in both Asia and North America, while the Ptarmigan *L. mutus* can be resident as far as 80°N in Svalbard and Greenland (Cramp and Simmons 1980, pp. 392 and 408). Both species feed almost entirely on plant material for which they may need to dig through the snow.

The Snowy Owl *Nyctea scandiaca* breeds in northern Greenland to within 6° of the north pole. However, it is not resident this far north and most of the population is partially migratory and nomadic during the winter months, but even then the majority of the population stay above 60°N. In Canada and Alaska it is argued that most birds probably withdraw from the high-arctic zone where 24-hour darkness can be experienced (Snyder 1957; Godfrey 1966). However, some birds are reported to be resident as far north as Germania Land in eastern Greenland between latitudes 74–80°N (Salomonsen 1950), where the sun stays continuously below the horizon from early November till mid February. What is perhaps most surprising about the Snowy Owl is that, when living at latitudes where more typical patterns of night and day occur, it is not usually nocturnal, preferring to hunt through the day and occasionally into the twilight.

The Gyrfalcon *Falco rusticolus* has a similar circumpolar breeding distribution to the Snowy Owl and like that bird breeds as far north as the very northern tip of Greenland (latitude 83°N). Although the most northerly populations are migratory there are still resident populations as far north as 70° where light levels below those of day-time are continuous for a number of weeks.

While these four species are birds primarily of high northern latitudes throughout their breeding range, one other bird which may winter there has a much larger latitudinal range. The Raven *Corvus corax* breeds from within 10° of the equator in Nicaragua to within 10° of the north pole in Greenland and on Ellesmere Island (Goodwin 1986, p. 125). Not surprisingly with a species of such wide latitudinal and climatic range, many subspecific differences have been identified, principally that birds from higher latitudes are larger. However, there seems to be no study of how these Ravens survive the arctic winter and cope with the long periods of darkness which they must experience.

Among factors which could help to explain the ability of all these birds to

Ravens *Corvus corax* are one of the few birds apparently able to survive during the long darkness of Arctic winters.

forage through the high-arctic winter are the following. First, extensive cloud cover at these latitudes is relatively rare and so when the moon is available light levels will nearly always fall within the higher moonlight range. Secondly, at these latitudes the aurora borealis can be frequent and very bright, at times providing levels of illumination close to and even exceeding those of full moonlight (Natural Illumination Charts 1952). Thirdly, because of its high reflectance, the luminance of the snow could be up to about ten times higher (1.0 log unit) than shown in Figure 2.5 for temperate habitats under the same conditions of natural illumination. Whether these factors either alone or together are sufficient to account for these birds' ability to be both mobile and able to locate prey is not clear.

There is, however, one factor which the habitats occupied by these birds have in common with those used by all of the species discussed in the preceding section, this is that the habitats are open and treeless, so that the birds are unlikely to have to negotiate small obstacles and make fine spatial discriminations when in flight. The Ptarmigan and Red Grouse forage on the ground, often digging into snow in order to unearth vegetation. The Gyrfalcon in winter preys principally upon these two species in the same open habitat but may also take small mammals which are also the main prey of the Snowy Owls. The falcon hunts principally by low-level flight over open terrain or by short flights from low observation points, the latter technique is also used by the owl. In general the Raven is less specialised in both diet and foraging technique than any of these birds. Whether those Ravens which survive the winter thus far north adopt specific or peculiar foraging strategies is clearly worth investigating; if anyone can bring themselves to do it!

NOCTURNAL AERIAL ROOSTING

Perhaps one of the most intriguing examples of occasional nocturnal behaviour is provided by the Swift *Apus apus*. Radar studies have shown that Swifts

The evening ascent of Swifts *Apus apus* to a high altitude where they seem to spend the night "roosting on the wing".

regularly "roost on the wing" at high altitude throughout the night (Lack 1956; Cramp 1985, p. 662). During the breeding season it is mainly the one-year-old non-breeding birds which roost on the wing whereas the breeding birds roost on walls, ledges or in cavities at or near their nest sites. Radar has revealed that at dusk, flocks of birds climb rapidly to a height of between 1,000 m and 2,000 m and occasionally even higher. They then fly around at these heights throughout the night, sometimes drifting with the wind. They descend to the area of the breeding colonies around dawn. Towards the end of the breeding season, the adult breeding birds may join in such aerial roosting and there is some evidence that birds may roost on the wing at night in their African winter quarters. Just why they should choose to roost on the wing throughout the night is not understood though the Swift is regarded as one of the most aerial of all birds and the energetic costs of such roosting flights may not be high. However, it is worth noting that, as in nocturnally migrating birds, all flight takes place within the open airspace well away from any obstacles. The birds neither enter nor leave the nest site during night-time but restrict all of their nocturnal activity to flight in open airspace.

CONCLUDING REMARKS

The purpose of this chapter has been to provide a broad introductory perspective of the activities and sensory problems of the truly nocturnal birds. The overall picture which emerges is that nocturnal activity in birds cannot be taken for granted.

This chapter has dwelt at some length with diverse examples of the nocturnal activities of several species of birds which on first consideration might be thought to have exclusively diurnal life styles. All of the species discussed here are those which we would expect to see going about their daily routine at midday. That these same birds could be engaged in purposeful activities at midnight is perhaps surprising. However, these night-time activities are all restricted in some way compared to those of the daylight hours. It is not simply that these birds have the necessary sensory capacities to allow them to indulge in their full range of activities regardless of the light level. Rather, as has been shown, certain sensory capacities coupled with a restricted behavioural repertoire, conducted in particular environments, can permit nocturnal activity.

The sensory capacities discussed here as important for nocturnal activities may have surprised some readers who see birds primarily as "a wing guided by an eye". Birds have at their disposal a range of sensory capacities which provide information about aspects of their environment which are both remote from them as well as in touch with their bodies. However, knowing what sensory information is available to a bird does not explain how it is used to guide behaviour. For example, there is much information on the range of sensory capacities which a bird may have at its disposal to determine the appropriate departure heading of its nocturnal migratory flight. However, just how they are used to guide the actual flight is not understood. We seem only to scratch the surface of such problems but knowledge of the possible sensory repertoire does give some idea of the problems and their possible solutions.

Perhaps the most important general points to have emerged from this discussion are:

First, it is not that certain species simply extend their day-time repertoire beyond dusk; it is rather that activity at night is somewhat restricted and presents special problems which require specific sensory and behavioural solutions.

Secondly, in all cases where flight is involved this takes place either in open air space or in open habitats largely devoid of small obstacles. Where birds are active in spatially complicated habitats, as are some night-singers and those petrels which nest under a tree cover, the birds tend not to fly.

Thirdly, occasional nocturnal activities are often not preferred to day-time ones. It may be energetic or metabolic considerations which result in the birds being active by night, rather than that they prefer to be so. The corollary is that although these birds can be active at night under certain circumstances they are not necessarily sensorily well equipped to do so. Their sensory capacities permit only certain restricted behaviours rather than the full repertoire which we might expect to find in truly nocturnal birds. This argument is probably most contentious in the case of nocturnal migration, but there is certainly evidence that some species employ a strategy of mixed nocturnal and diurnal migration, choosing to travel at night only where energetic considerations force them to do so. It was also seen that a similar argument applied to occasional night feeding among the waders and water-fowl. For them it seems likely that foraging and feeding methods are not dependent upon visual guidance. Their reliance upon other sensory cues may predispose them to exend their feeding into night-time when food require-ments demand. Feeding at night-time light levels in the non-migratory populations of some arctic birds in winter, especially the Raven, remains an enigma. Why these birds choose to stay north of the arctic circle during the prolonged darkness of winter, and how they cope with the sensory and other problems that this would seem to imply, is not clear.

These examples of species which engage in occasional nocturnal activity are not, of course, exhaustive. They have been drawn in the main from species found in Europe and North America. However, it seems likely that the principles and factors which underlie these different nocturnal activities are common to most other examples of occasional nocturnal activities in birds.

CHAPTER 5

Crepuscular activity in birds

DEFINING TWILIGHT AND CREPUSCULAR ACTIVITY

The adjective "crepuscular" means simply "of twilight". For humans twilight, especially at the end of the day, is regarded as a particularly trying period. The difficulty arises because at this time we often try to continue day-time activities even though our ability to see detail and discriminate between colours is decreasing. In legal terms the lower limit of civil twilight, evening or morning, defines the time when outdoor work requiring daylight can begin or must end. In literature "twilight" is a cliché for mystery and difficulty, a time when things are not quite what they first seem. It is a time when activities begin or end rather than when activities proceed in their own right. But how should twilight and its associated crepuscular activities be regarded in the natural world?

As was seen in Chapter 2, twilight embraces a wide range of constantly changing light levels and for convenience is divided into three sections; civil, nautical and astronomical twilight. Strictly, twilight refers to all of those times when the sun is below the horizon but the influence of its light (other than that reflected from the moon) can still be detected. This extends from the time of sunset right through to the end of astronomical twilight. The range of light levels which this embraces clearly overlaps with the light levels which can be experienced in the middle of the night due to starlight and/or moonlight (Tables 2.1 and 2.2).

A further problem with twilight is that its duration can vary markedly with season and latitude. Thus, in midsummer even at moderately high latitudes, such as 60°N (that of Shetland, Oslo, Stockholm, Leningrad, Anchorage, or the southern tip of Greenland), civil twilight may last over three and a half hours each day. Even at the time of the equinoxes when civil twilight is shortest, it still constitutes nearly one and half hours of the day length at this latitude. At the equator, however, civil twilight lasts approximately 40 minutes every day and is uninfluenced by the time of year. Thus, depending upon its latitude and on the time of year, a crepuscularly active bird will have quite different lengths of time in which to conduct its activities each day.

The actual light levels in these periods of twilight are changing over a large range. Civil twilight (without taking into account the effects of cloud cover) spans a range of about 100-fold, while nautical twilight embraces light levels which may vary by 10,000-fold. During civil twilight at the equator, light levels fall by a factor of 10-fold every two minutes, while at latitude 60° the rate of change of light levels during the same twilight period is five times slower than this.

Strictly, twilight begins or ends at the *time* of sunset or sunrise. Light levels above those produced at this time, when the sun's rays can be experienced directly, are regarded as daylight. But, when discussing the crepuscular activity of animals, by what should the lower limit of twilight be defined? When considering a definition of nocturnality in Chapter 3, a distinction was made between a reasonable definition of night-time light levels and a definition which could have some use as regards field observations of an animal's activity.

In order to provide continuity with those definitions the following limits of twilight are suggested and are based upon summarised data of Figure 2.5. *Twilight refers to the full range of light levels which may be experienced between sun rise/set and that of the beginning/end of civil twilight under maximum cloud cover.* This definition applies to the situation in open habitats where luminance levels of the substrate (leaf litter, grass, etc) vary between those experienced at sunset under a clear sky (or for some time before sunset under overcast), to that which can be experienced under maximum moonlight without cloud. This latter light level was used in Chapter 3 to define the upper limit of night-time light levels and thus provides continuity between the two light ranges. Thus, according to this definition, any activity which takes place within the 1,000-fold range of light levels defined above should be regarded as crepuscular activity. The disadvantage of this definition is that the lower limit of twilight is not readily observable in the field without light measuring equipment, and it is hence unlikely to correspond with criteria used to guide observations of crepuscular activity which find their way into field guides and handbooks.

In the case of nocturnal activity an attempt was made in Chapter 2 to provide a definition of strict nocturnality. A similar definition for strict crepuscularity is not attempted since it seems unlikely that any species actually falls into such a category, rather it is necessary to talk only of the *crepuscular activities* of birds rather than of a strict *crepuscular habit*. Viewing crepuscularity in this way takes account of the fact that many birds which are best regarded as either diurnal or nocturnal during most of their life cycle, may either start or end their period of daily activity during twilight and thus may be regarded as crepuscularly active. It also allows for the fact that many birds may be only occasionally active during twilight. For example, species such as the waders and waterfowl discussed in the previous chapter may, when on their wintering grounds, continue feeding from daylight, through the twilight period into night-time, but on other occasions feed only during the daylight period. On the other hand some of these same birds when on their high latitude breeding grounds may experience very long periods of twilight (Figures 2.3 and 2.4) during which they continue to forage and carry out their breeding activities.

CREPUSCULAR ACTIVITIES IN BIRDS

A cursory look through field guides and handbooks would suggest that crepuscular activity in birds is rather uncommon. Only occasionally does activity at dusk ever receive a specific mention and, when it does so, this is usually to record foraging behaviour, especially in species such as the owls and nightjars which may then continue to forage into night-time. However, such an emphasis on foraging, while clearly of importance, is misleading since very many bird species engage in other behaviours during the twilight period.

The following discussion presents just four examples of different types of crepuscular activities in birds. While this cannot represent a comprehensive coverage of the activities of birds at this time, these examples do illustrate certain general points about the ways in which twilight light levels may influence behaviour. The principal stumbling block for a more comprehensive discussion is that only rarely have light levels during the twilight activities of birds been specified and recorded.

The four examples (the dawn chorus of passerines, display flights of woodcocks, dusk hunting in falcons and the Bat Hawk *Machaerhamphus alcinus,* and dusk feeding in skimmers) are of birds which are generally regarded as diurnal in their habits, but which conduct certain of their activities in twilight. There are of course many other examples of occasional twilight activity in diurnal birds, but these are generally not concerned with specific tasks and are perhaps best regarded as preparations for entering or leaving a roost, regardless of whether the birds are primarily nocturnal or diurnal in habit.

It might be thought appropriate to consider here the twilight foraging of nocturnally active birds such as the nightjars (Caprimulgiformes) but since such activity usually represents the beginning of foraging behaviour which will continue well after dusk, it is more appropriate to consider it later under the heading of truly nocturnal behaviour.

DAWN CHORUS AND THE DAILY ONSET OF SONG

Perhaps the most widespread, and certainly the most familiar, crepuscular activities in birds are the dusk and dawn choruses, in which perhaps all passerines take part in varying degrees. The dawn chorus proper is a phenomenon of the breeding season in temperate latitudes, but even outside the breeding season the onset of song may occur during twilight. Investigations of the timing of the dawn chorus and of the onset of song at other times of the day, have shown that a complex of environmental factors may be involved. These include temperature, wind, humidity, barometric pressure, rain and light level (Armstrong 1963). Furthermore, it is now clear that these environmental factors interact with each other, with circadian rhythms of activity and with the bodily condition of individual birds and pairs. It seems that it is this complex of interactions which determines the timing and length of the dawn chorus. The function of the dawn chorus is still a matter of disagreement. The current debate centres on the duration and strength of the male's song during the chorus as a possible indicator of both bodily or phenotypic condition, and as a form of mate guarding at a time when female fertility is particularly high (Kacelnik and Krebs 1982; McNamara *et al* 1987; Mace 1987a,b; Cuthill and Macdonald 1990).

Early investigations of the factors which initiate the dawn chorus (e.g. Haecker 1916) concluded that it was specific light levels which triggered a bird to start singing. This work was supported by a number of subsequent studies (e.g. Doring 1920; Dorno 1924), all of which recorded that different passerine species began singing at different light intensities during twilight; they also showed that the onset of song could be retarded by the presence of cloud cover. However, none of these studies actually measured the light levels at which birds began to sing but relied instead upon time of day and estimates of cloud cover. As pointed out in the introduction to this chapter such estimates could lead to considerable errors depending upon the time of year and latitude.

When light levels were measured for the first time (Schwan 1920) the results were not so clear cut and Lutz (1931) suggested that, "light has little or nothing to do with the time of the first song, for light was, on the average, actually less when the bird sang early than when it sang late". Leopold and Eynon (1961) went on to show that the situation is in fact more complicated. They demonstrated that the onset of song in a range of species became increasingly earlier relative to sunrise as spring progressed, although song onset was retarded on cloudy days. This suggests that light levels were important although the actual level which triggered song changed through the season.

At about the same time that these studies were being conducted, understanding of the various ways in which the activity of birds is controlled by so-called biological clocks or circadian rhythms was increasing (Aschoff 1967; Gwinner 1975) and Schmitz and Middel (1966) were the first to propose that such rhythms may underlie the timing of daily song onset in Chaffinch *Fringilla coelebs*, Greenfinch *Carduelis chloris* and Great Tit *Parus major*.

Astrom (1976), in a more detailed study, investigated the interaction of light level, temperature and humidity on the onset of dawn chorus song in the Reed Bunting *Emberiza schoeniclus*. This study also suggested that the start of singing is in fact entrained by an internal clock correlated with the time of local sunrise as it changes from day-to-day, not by specific light levels, temperature or humidity. Indeed, Astrom concluded that all birds with a pronounced dawn chorus, "must be regarded as 'clocks' rather than 'photometers' [light measuring devices], at least in the morning". It is well established (Gwinner 1975) that the internal clocks of passerine birds can be entrained precisely by various external "Zeitgebers" (time-setters), the most important of which are daily changes in the duration of sunlight and twilight (Aschoff *et al* 1970). Thus, passerines involved in morning twilight song are thought to be predicting the time when they should start to sing rather than responding to the actual light levels produced during a particular stage of twilight.

That the overall activity patterns of birds may be responding to some kind of circadian rule, rather than actual light levels or some other environmental factor such as temperature, has also been suggested as an explanation of the roosting behaviour of birds during the evening twilight. A study of the Brown-headed Cowbird *Molothus ater* by Krantz and Gauthreaux (1975) concluded that the principal factor which determined when these birds entered their communal roost was not the ambient light level but the total amount of solar radiation to which the birds had been exposed during the day. In essence, the brighter the day overall (more solar radiation) the later the birds started to roost in the evening.

Astrom recorded much individual and day-to-day variation in the time of song onset but he also noted that Reed Buntings started to sing earlier and earlier relative to sunrise up to mid-June (the summer solstice) and later and later after that time. The consequence of these changes was that in spring the birds started to sing at lower and lower illumination levels up until midsummer, with the reverse pattern thereafter. A similar pattern has also been found in the Yellowhammer *Emberiza citrinella* (Blase 1971). In both of the above birds the actual range of light levels associated with song onset lay between that equivalent to maximum moonlight and the beginning of civil twilight. That is, the birds had all started their first songs when light levels were between 100 and 1,000 times lower than that of sunrise.

As discussed in Chapter 4 singing at night does not require birds to be particularly mobile, and indeed in the case of true night singing there is evidence that birds do remain in one place and sing from regular perches. There are reports that this also tends to be true of the Reed Bunting, Great Tit and Blackbird, at least for the early part of the dawn chorus (Astrom 1976; Mace 1986; Cuthill and Macdonald 1990). It is not surprising, therefore, that the start of this activity is somewhat independent of the actual light levels during twilight and can be influenced by such factors as the bodily condition of the male and the fertility of its mate (Mace 1986; Cuthill and Macdonald 1990).

But when does the dawn chorus end? Is it when birds start to leave their song perches and begin to forage? Important information on this question has come from studies of foraging and twilight light levels in the Great Tit and of how female Great Tits influence the length of the dawn chorus singing of their mates.

By combining laboratory studies of foraging efficiency and measures of the actual luminance levels of prey items in the field, Kacelnik (1979) showed that Great Tits would not begin foraging for food items until light levels were above those experienced in the bird's natural habitat after sunrise. Furthermore it was shown that searching efficiency in Great Tits continued to improve with increasing light levels and did not reach an asymptote until light levels had increased 1,000-fold above those of the initial foraging level. It was also shown that in the Great Tit's natural broad-leaved woodland habitat this did not occur until between one-and-a-half to two hours after foraging had begun (i.e. until after sunrise). Direct observation indicated that the foraging birds were guided by visual cues. Thus it would seem that in this particular species of small passerine, only relatively simple non-visually guided behaviour is possible during twilight. Only after sunrise has been reached can more complex visually-guided behaviour such as foraging be undertaken, and even this is initially at a low level of efficiency.

Kacelnik (1979) proposed that passerines sing, rather than forage, at the time of the dawn chorus partly because efficient visually-guided foraging is not possible at twilight luminance levels. It seems clear that foraging and singing are competing activities in Great Tits even at high day-time light levels (Mace 1989). That a small passerine bird might be expected to forage as early as possible is suggested by the fact that overnight they may lose 5–10% of their body weight (Baldwin and Kendeigh 1938; Owen 1954).

It has been shown that during that part of the breeding season when eggs are laid, the time when a paired male Great Tit stops singing is determined by the emergence of his mate from her roost (Mace 1986). Following her emergence the pair copulate, presumably because of her high fertility at this particular time of day. Hence, in this case the female's behaviour controls the duration of her mate's dawn chorus. However, it is not known what factors control the female's emergence time but it seems possible that she does not emerge until foraging can begin. Thus she usually emerges after the male has been singing for at least 20 minutes and after the beginning of civil twilight, i.e. during the period when Kacelnik showed that light levels were likely to be sufficient for successful foraging to begin.

The generality of these findings for understanding how foraging in passerines is affected by light levels is not clear. These studies refer specifically to birds which spend about half of their foraging time on the floor of woodlands (Gibb 1954; Perrins 1979), where light levels are likely to be low compared with conditions outside the wood. However, it would seem safe to conclude that in many passerine species, efficient foraging is not possible until light levels at least reach those of between civil twilight and sunrise and that whatever the function of the dawn chorus, most birds are not able to forage efficiently during this time.

DISPLAY FLIGHTS OF WOODCOCKS

In Chapter 4 it was noted that Eurasian Woodcocks (and possibly American Woodcocks) forage at night outside woodlands. This nocturnal activity is only occasional since it occurs mainly outside of the breeding season. When breeding, the birds forage principally by day on the woodland floor. These same species also provide examples of birds whose breeding displays are performed almost exclusively during twilight.

The displays of the Eurasian and American Woodcocks *Scolopax rusticola* and *S. minor* have been studied in some detail (Sheldon 1967; Cramp and Simmons 1983), although many aspects of this behaviour are not fully understood. It is thought that the behaviour of the other four species of the *Scolopax* genus are similar to that of the Eurasian Woodcock but very little is known about any of these species, which are confined to small populations on islands (Hayman *et al* 1986).

The crepuscular display of the male Eurasian Woodcock can be observed throughout their protracted breeding season. It is so conspicuous and unique among British birds that a vernacular name, Roding, has been attached to it, possibly since Anglo-Saxon times. Roding has now become a scientific term.

Roding occurs at both dusk and dawn though it is more frequent and sustained at dusk. It is only rarely recorded during the day and then typically at high latitudes where nights and breeding season are short. Roding birds fly slowly above the tree canopy moving the wings in a characteristic exaggerated manner while regularly delivering a song made up of various croaking, growling and sneezing sounds. The first flight of the evening may be high above the canopy, becoming lower and slower as light levels fall. Each display flight may last up to 20 minutes but the average is about six minutes. The function of the display is self-advertisement with the males trying to reveal their presence to receptive females on the floor of the wood beneath (the mating system is polygynous or promiscuous). The female may call down a roding male using so-called "sneeze notes", or fly up from the wood and fly alongside the male before the pair land together.

The self-advertising display of the American Woodcock also occurs at dusk and dawn throughout the breeding season but differs from that of the Eurasian Woodcock in that males display from display stations on bare ground in woodland clearings or at woodland edges. The displaying bird

The Woodcock *Scolopax rusticola*, one of the few species which usually restricts its courtship display flights to twilight periods.

adopts a characteristic posture while uttering a series of calls. At intervals of a few minutes the bird rises vertically with a characteristic flight pattern described as on "twittering wings" and delivers a song whilst giving a circular display flight about 100 m above the ground. After a few minutes it returns to the ground in a rapid spiralling flight. Thus, while the Eurasian male flies over the woodland waiting to be attracted down by any receptive female that it passes over, the display of the American bird serves to attract females to its display station.

Why woodcock should chose twilight in which to execute these display behaviours is not clear. However, in all species both sexes are particularly secretive and solitary. Furthermore, due to their superb cryptic coloration they are very difficult to observe on the woodland floor where they spend most of their time. By engaging in stereotyped flight activity above the woodland canopy the conspicuousness of a displaying male, especially to a female viewing in silhouette from below, is likely to be considerably enhanced. More importantly perhaps, this enhanced conspicuousness occurs at a time when most other woodland birds, including potential predators such as the Goshawk *Accipiter gentilis,* have gone to roost. In both species the females can remain inconspicuous and do not have to fly during the pairing process.

While the visibility of the male woodcocks is enhanced by viewing their flight in silhouette against the evening sky it seems not unreasonable that the male is not able readily to detect or identify the female below until she calls or visits the display station. Thus there are reports that roding Eurasian Wood-cocks may fly down and approach, as though towards a female, other birds flying at dusk. These may range in size from a Starling *Sturnus vulgaris* to a Marsh Harrier *Circus aeruginosus.* Furthermore it has long been known that roding birds can be attracted to dummies and decoys and a range of such devices have traditionally been used to trap birds in Europe. Their efficacy attests to the poor visual discrimination of the woodcock whilst roding during twilight. As in the case of other nocturnal and crepuscular flights discussed so far, all of the twilight flight activity of the woodcocks takes place in the open, well away from vegetation.

CREPUSCULAR HUNTING IN FALCONS

Although the twilight hunting of falcons has been recorded many times it has never received the kind of detailed study, described above, concerning the crepuscular foraging and singing of passerines. There seem to be no studies of the light levels at which this hunting takes place, and most data is in the form of anecdotal observations in handbooks or species monographs, see, for example, Brown and Amadon (1968), Brown (1976a, p. 227). It would seem that crepuscular hunting is at most an occasional activity within the context of the annual cycle, and the circumstances under which it occurs have not been studied. The species involved are mainly the smaller species of the genus *Falco.*

In the Western Palearctic region, among the species recorded as occasion-ally hunting at twilight are: Lesser Kestrel *F. naumanni,* Kestrel *F. tinnuncu-lus,* Hobby *F. subbuteo,* Sooty Falcon *F. concolor* and Eleonora's Falcon *F. eleonorae.* Entries for each of these species in Cramp and Simmons (1980) record either hunting in twilight or hunting at night in the light from artificial sources, such as street lamps or floodlights.

It seems likely that smaller *Falco* species in all parts of the world may be occasional twilight-hunters. For example, in the Australasian region both the Brown Falcon *F. berigora* and the Australian Hobby *F. longipennis* have been observed hunting occasionally in twilight (Mooney 1982; Price-Jones 1983). A less well known species of falcon from Central and South America, the Bat Falcon *F. rufigularis,* as its name implies is noted for taking bats at dusk, though its reliance upon this prey may not be exclusive.

Eleonora's Falcons *Falco eleonorae* continue to take large insects well beyond dusk.

There are also reports of occasional twilight-hunting in some of the larger *Falco* species, such as the Peregrine *F. peregrinus* (Ratcliffe 1980), and even reports of an apparently exceptional individual Peregrine which specialized in "nocturnal" hunting, with the aid of artificial lighting (by which it was observed), which took birds over the waters of a harbour at night (Clunie 1976). Presumably, those Gyrfalcons *F. rusticolus* which winter far north must hunt at twilight light levels at certain times of the year (see section on Arctic birds in Chapter 4).

To gain some understanding of the crepuscular activities of all these species it is perhaps sufficient to consider in detail only the few Western Palearctic species listed above. In addition to hunting at dusk, the Hobby, Eleonora's Falcon and Kestrel have been recorded hunting by moonlight, while Lesser Kestrels and Sooty Falcons have been observed hunting in the light from flood lamps (Cramp and Simmons 1980). All of these falcons feed by taking insects, small birds and occasionally bats, on the wing in the open airspace; the Kestrel also takes small mammals from the ground. Through most of the Kestrel's range, diurnally active voles are an important element of its diet, but in areas where these are absent (Ireland and Île d'Ouessant, France) the Kestrel is reported to rely on normally nocturnally active wood mice (Thiollay 1963; Fairly 1973; cited in Cramp and Simmons 1980), though exactly when these are caught is not recorded.

These falcons share with the occasionally nocturnal species discussed above (waders, waterfowl, sea-birds) a preference for foraging in open habitats usually devoid even of large obstacles. The Sooty Falcon hunts principally over treeless lowlands, unvegetated deserts and bare marine islands, while Eleanora's Falcon hunts mainly over the seashore or open sea, and the two kestrel species and Hobby prefer short grasslands and dry, lightly vegetated, areas. At first sight, the hunting techniques of these birds would seem to require a relatively high degree of visual guidance, since it often involves taking relatively large prey items such as small passerine birds, mainly with the feet rather than "trawling" prey from the air space. Nevertheless, Sooty and Eleanora's Falcons, which are the most regularly crepuscular of the group, and to a lesser extent Hobby and Lesser Kestrel, all depend on insect species which occur in large aerial concentrations, such as locust, dragonflies, flying ants and termites (Brown and Amadon 1968; Cramp and Simmons 1980). It is upon such insect concentrations that the birds feed when exploiting artificial light sources. When feeding on insect swarms, prey is sometimes taken "trawling fashion" in the bill, rather than with the feet, suggesting that visual guidance to individual prey items is then not employed. That the birds hunt at night, apparently only when bright moonlight or artificial lighting is available, also suggests that vision is a limiting factor and it would be of interest to know whether crepuscular hunting is seasonal (there is evidence that passerine birds form a significant proportion of these falcons' diet only during the breeding season), at what stages of dawn and dusk twilight hunting actually takes place, and the level of foraging efficiency achieved during twilight compared with day-time. Various reports do suggest, however, that the efficiency of prey capture (especially involving small passerine birds) in these and other species of falcon, even during daylight, is relatively low (Brown and Amadon 1968; Walter 1979; Ratcliffe 1980).

CREPUSCULAR HUNTING IN THE BAT HAWK

One further bird of prey which does deserve mention in this context is the Bat Hawk *Machaerhamphus alcinus*. It is a relatively rare bird of wide distribution, whose natural history is only poorly understood (Brown 1976a; Newton 1985), but of all birds it may come closest to being a specialist crepuscular feeder. As its name suggests this bird is noted for preying upon bats as they emerge from their roosts at dusk. It inhabits tropical latitudes (Africa and the Far East) where the twilight period is particularly short, but the bird has the advantage that twilight and prey are both highly predictable from day-to-day, and at many feeding sites prey may be extremely abundant. One particularly dramatic example is the Niah Caves complex in Borneo, where Bat Hawks can regularly be observed hunting at dusk. In these caves it is estimated that four-and-a-half million swiftlets breed alongside many

One of the few specialist crepuscular feeders, the Bat Hawk *Machaerhamphus alcinus*, takes bats as they emerge from their roost at dusk.

millions of bats (Harrison 1976). The former return in large numbers to the caves at dusk and the bats emerge at about the same time. Although a member of the Accipitriformes, the general appearance of the Bat Hawk is similar to that of a falcon and it is reputed to be capable of very swift flight. Unlike the majority of falcons and hawks, however, the Bat Hawk tends to take prey both with its talons and often, directly, with its wide gape.

Like the falcons mentioned above, the Bat Hawk does have a similar requirement for the open-habitat types in which it hunts, and the bird will make use of wide rivers, lakes and short grasslands, including lawns. There are reports of it hunting at night on bats attracted by insects which in turn are attracted by artificial lights (Brown 1976, p. 69). As with the crepuscularly active falcons, it would clearly be of great interest to gain further data on the Bat Hawk's hunting techniques and on the circumstances of its crepuscular hunting since, as already mentioned, this bird could be one of the few specialist crepuscular predators among birds. However, even this bird also feeds during the day on species of small colonially breeding birds, notable Hirundinidae (martins and swallows) and Apodidae (swifts and swiftlets).

CREPUSCULAR FORAGING IN SKIMMERS

The skimmers are a fascinating group of three closely related species whose specialised feeding technique was referred to briefly in Chapter 4. The three species show many similarities and are sufficiently specialised to have been allotted to the same genus (*Rynchops*) within their own family (the Rynchopidae) in the Charadriiformes. The three species replace each other around the

Whether feeding by night or by day, Black Skimmers *Rynchops niger* probably use touch sensitivity in their lower mandible to locate prey items.

globe in tropical and sub-tropical regions and have been of much taxonomic interest, and are now regarded as comprising a superspecies (Sears *et al* 1976).

The most well studied species is the Black Skimmer *R. niger* from the Americas. In particular, its feeding technique has been studied in considerable detail (Erwin 1977; Zusi 1962). Though these birds feed by day there are many records of them feeding during twilight, and even at night (Zusi and Bridge 1981, 1985). It seems clear, however, that it is the use of open habitats without obstructions, and the sensory basis of its specialised feeding technique, which permits this bird to feed at these times.

As discussed in Chapter 4, these birds feed by flying in straight, level flight over shallow calm waters (wide rivers, lagoon, lakes) with the lower mandible trailing just below the water surface. Tactile cues, produced when the mandible strikes a fish, trigger a fast reflex action whereby the bill is snapped shut and the fish caught whilst the bird continues its flight. To achieve this the skimmers not only exhibit special adaptations of the bill (the lower mandible is longer than the upper mandible and is blade-shaped in cross secton), but the neck vertebrae and musculature also show a number of specialisations to absorb the force of an impact with an object and to flex the neck rapidly when retrieving prey whilst still maintaining level flight (Zusi 1962). It is also known that the birds choose to feed with this technique over shallow waters where small fish occur at particularly high concentrations. Thus, the skimmer's feeding technique does not require the visual detection of individual prey items. Rather, the birds trawl through shoals of fish relying upon tactile cues for prey detection and capture. Zusi (1985) suggests that the birds are attracted to suitable foraging areas by the disturbance of otherwise calm surface waters produced by shoals of fish. Presumably the suitability of individual prey items caught in this way is determined by tactile and taste cues within the mouth.

The skimmers' feeding technique is not without problems. The birds may strike underwater objects or drive into the muddy bottom of a lagoon, and breakage of the tip of the lower mandible is not uncommon. Nothing is known of the visual capacities of the skimmers. However, unlike the case of the nocturnally active Swallow-tailed Gull, which has a relatively large eye (Chapter 4), the eye of the Black Skimmer is reported to be relatively smaller than that of the diurnally active Common Tern *Sterna hirundo*, which forages in similar situations (Zusi and Bridge 1981). Skimmers are the only birds whose pupil is in the form of a vertical slit rather than circular. The visual significance of this is unknown but a slit pupil may be closed to a smaller aperture than a circular one and thus protect the eye from bright sunlight during the day (Zusi and Bridge 1981). Why other birds, such as the terns which forage alongside the skimmers, do not need similar protection is not clear.

COMMON CONSTRAINTS AND SOLUTIONS TO CREPUSCULAR ACTIVITY

The above discussion has centred upon just four specific instances of crepuscular activity. Deliberately excluded from the discussion have been activities which may occur during twilight but which also frequently extend into the night. These examples of specific crepuscular activity can be examined from two main perspectives. In the first instance there are activities associated with breeding (the passerine dawn chorus and the display flights of woodcocks) as opposed to foraging (falcons, skimmers and Bat Hawks). The second perspective is that only the passerine dawn chorus does not involve conspicuous flight, at least during its initial stages. However, the influence of the rapidly changing light levels on behaviour are understood only in the case of the dawn chorus.

In passerines, although the actual start of singing seems to be independent of light level during twilight, it seems likely that light levels do in fact exert a strong influence on the dawn chorus, albeit in a somewhat negative way. Thus, it seems that not until illumination reaches a certain level is efficient, visually guided, foraging possible. The consequence is that the dawn chorus can at least be viewed in part as resulting from the bird's inability to do little else other than sing during the early stages of dawn.

The influence of light levels on the crepuscular flights of the falcons, skimmers, woodcocks and the Bat Hawk have not been subject to systematic investigation. Nevertheless it is clear that in all instances the birds make use of open airspace for their activities. The actual prey catching techniques used by twilight hunting falcons are not clear, but they are not taking prey from the ground, or from amongst vegetation. The birds may be using visual pursuit for these aerial prey items which would be more readily seen viewed in silhouette against the sky. It is also possible that these birds may "trawl" insects or larger prey by flying through an aggregation of relatively slow moving prey with an open bill, rather than pursuing individual items.

In the case of the skimmers, the parallels between their foraging techniques and those of the nocturnally foraging shorebirds are clear. Although skimmers feed in flight while shorebirds feed on the ground, both feed within open habitats practically devoid of obstructions. Having identified a suitable feeding area where prey is abundant, both groups of birds normally forage (even during day-time) more or less blindly for individual items, detecting them by tactile cues from the bill. Thus, in the skimmers, it can clearly be seen that their foraging technique predisposes them to activity during twilight.

CHAPTER 6

Cave dwelling birds

There are a few bird species which are able to fly within the totally dark interiors of caves. They enter caves not for feeding but to roost and to nest and when so doing may fly hundreds of metres in total darkness. Clearly, how these birds are able to achieve this is of considerable interest to the theme of this book, although only one of the species is nocturnal. Not surprisingly a cave dwelling life style has aroused considerable curiosity amongst ornithologists, and many aspects of the natural history and sensory capacities of these birds have been studied, but many topics still require investigation.

94

THE CAVE BIRDS

The Oilbird *Steatornis caripensis* has a local distribution in northern South America and on Trinidad in the Caribbean. It exhibits so many unique features, both in its anatomy and natural history, that its status has engendered considerable debate among taxonomists since it was first discovered almost 200 years ago. At present the bird is regarded as the sole member of the family Steatornithidae within the Caprimulgiformes (nightjars, frogmouths, etc). Its breeding biology and activity within caves has been studied in some detail. The first accounts were published by the explorer Humboldt in 1817 and the most comprehensive accounts are those of Snow (1961; 1962). Its activity outside the nesting caves is still poorly understood.

The other group of birds to fly in the complete darkness of caves is the swiftlets, whose taxonomic status has also been the subject of much revision (Brooke 1970; Medway and Pye 1977). The exact number of genera and species has long been in dispute but the present consensus is that the group consists of 15 species divided between three genera in the Apodidae (Apodiformes); some species can only be separated by detailed examination in the hand. All of the species live within the tropical regions from the Indian Ocean, through the Far East, northern Australia and western Pacific Ocean. Not all species of swiftlet fly within the totally dark interior of caves; some populations and species nest and roost near cave entrances while others make use of buildings; and one species, the Giant Swiftlet *Hydrochous gigas*, frequently nests behind waterfalls (Medway 1962a; Becking 1971; Langham 1980). The species which are thought to penetrate furthest into caves are the Black-nest Swiftlet *Aerodramus maximus* and the Edible-nest Swiftlet *A. fuciphagus* though there is one record of the latter species breeding within a cave-like building (Langham 1980; Becking 1971).

Although Oilbirds and swiftlets use caves for breeding and roosting throughout the year they differ in important aspects of their ecology and natural history. First, only the Oilbird is nocturnal. The swiftlets are diurnal, coming to roost in the caves at dusk, with most birds returning around sunset and during the period of civil twilight, and leaving the roost mainly after sunrise (Medway 1962a,b; Langham 1980; Langham, personal communication 1988). It is when returning to the roost that the swiftlets may fall prey to the crepuscularly hunting Bat Hawk, referred to in the previous chapter. Secondly, swiftlets feed exclusively on aerial insects trawled in flight from the open airspace, whereas the Oilbird feeds only on fruits taken from treetops whilst hovering (Tannenbaum and Wrege 1984). It is the only bird species which feeds nocturnally on a fruit diet.

Despite their differences in natural history these two groups of cave dwelling birds both use "echolocation" to orient their flight within the dark interior of caves. Amongst birds this sensory capacity appears to be found uniquely in these species. There is also some evidence that the Oilbird may employ olfaction to aid its orientation within the nesting cave and also to find fruit when foraging. Tactile cues are probably important in guiding Oilbird behaviour about the nesting ledge and towards other birds after landing in the cave.

ECHOLOCATION AND CAVE DWELLING

Field observations and experimental studies show that both the Oilbird and the cave swiftlets can rely exclusively upon hearing for their flight orientation within caves. All of these species employ echolocation, or active sonar (SONAR is an acronym for SOund NAvigation Ranging). That is, they have the ability to detect the presence of objects by emitting sounds and analysing the echoes from those objects (Pye 1980). In contrast, passive sonar determines the position of an object from the sounds emitted by the object. Passive sonar is the basis of all other forms of auditory localisation used by birds.

Not all species of swiftlet have been shown capable of echolocation, the exceptions being the Giant Swiftlet and the Glossy or White-bellied Swiftlet *Collocalia esculenta* (Medway and Pye 1977) which do not nest deep in caves. There is a report that the Humboldt's Penguin *Spheniscus humboldti* is able to detect both live and dead fish using active sonar (Poulter 1969) but there is no evidence that any other bird possesses the ability to echolocate as a means either of orientation or prey detection (Pye 1985). This may seem surprising in view of the specialised nocturnal activities of some other bird species discussed later in this book. It is clear, however, that the echolocatory abilities of swiftlets and Oilbirds are relatively crude and suited only for use within the spatially simple and predictable interiors of caves. Indeed, Snow (1961) suggests that echolocation is used exclusively by Oilbirds within their caves and is not used to orient the birds when outside the caves at night. However, Knudsen and Konishi (1979) do comment that Oilbirds may occasionally emit the kinds of sounds employed in echolocation when outside the cave. There is no evidence that swiftlets ever use echolocation outside the caves.

Certainly both Oilbirds and cave swiftlets are incapable of detecting and discriminating the fine spatial details that are achieved with echolocation by many aerial mammals, such as Microchiropteran bats (Pye 1979), or by aquatic mammals, such as dolphins (Johnson 1986). The echolocatory abilities both of swiftlets (Novick 1959; Medway 1959; Griffin and Suthers 1970; Griffin and Thompson 1982; Fenton 1975; Smythe and Roberts 1983) and Oilbirds (Griffin 1954, 1958; Konishi and Knudsen 1979) have been the subject of experimental investigations, and neither group of birds has been shown capable of employing echolocation to detect prey.

Experiments designed to determine the smallest objects which swiftlets can detect in flight have produced contradictory results (Smythe and Roberts 1983). The smallest targets reported to be detected by various species of swiftlets were wooden rods or wires with diameters of 4–10 mm. In the Mountain Swiftlet *Collocalia hirundinacea = Aerodramus hirundinaceus* there was some evidence that wires as thin as 1.5 mm could sometimes be detected (Fenton 1975), but in the Grey Swiftlet *Aerodramus spodiopygius* Smythe and Roberts (1983) concluded that the threshold for detection lay between 10 and 20 mm diameter. Griffin and Thompson (1982) put the threshold of *Aerodramus spodiopygius* at about 6.3 mm. Clearly these performances are not particularly impressive, especially when compared with what bats can achieve in comparable tests when wires as small as 0.1 mm diameter can be detected (Bradbury 1970; Simmons *et al* 1975). Smythe and Roberts (1983) report that swiftlets occasionally collide with thin stalactites even within familiar caves.

It should be noted that these tests simply involved investigating the size of objects which the birds could avoid as they flew towards them. Nothing is known about the distance, shape and volume of the space around the bird where such targets can be resolved. While sound production by the birds is likely to be somewhat directional, their hearing appears not to be well focused as it is in many bats (Simmons 1969). Such focusing, which the bats achieve by their large pinnae (outer ears), reduces the chances of interference from the calls of other bats flying in the vicinity. The Oilbird seems to have no mechanism to avoid such interference (Konishi and Knudsen 1979). Also, the distance over which these echolocatory mechanisms can yield spatial information will be limited by the volume of sound that the bird produces, and also by the presence of other sounds, principally those produced by other birds, within the cave. By all accounts the interior of these nesting caves is noisy due to the calls of birds flying within them (Snow 1961; Medway 1962a). Suthers and Hector (1985) reported that the volume of echolocatory clicks and squawk-like vocalisations produced by an Oilbird typically exceed 100 decibels at a distance of one metre.

The minimum size of objects which the Oilbird can detect using echolocation is even larger than those detected by swiftlets. When plastic discs of various diameters were hung in the passageway leading out of a nesting cave, "All birds hit 5 and 10 cm diameter discs, as if nothing had existed in their paths. The first sign of avoidance appeared when 20 cm discs were presented and all birds avoided the 40 cm discs." (Konishi and Knudsen 1979, p. 426.) For an animal which has achieved renown as an "echolocatory" species this performance is surprisingly poor and clearly indicates that echolocation can only be used to guide the birds with respect to large objects such as the walls of the cave or other Oilbirds. What is perhaps more interesting is that Oilbirds choose to fly in complete darkness when they are clearly incapable of making fine spatial discriminations to guide their flight. Perhaps they will do so only in the spatially simple and well known interior of their nesting caves.

ECHOLOCATION IN HUMANS

An interesting context in which to view the Oilbird's echolocatory performance is to note that blind humans are capable of at least comparable, if not superior, echolocatory performance (Kellogg 1962; Ammons *et al* 1953). In the case of blind humans the sound source used may be repetitive clicks of the tongue, short words or taps, while the Oilbird and swiftlets emit trains of short sound pulses, perceived by observers as a train of clicks. In the Oilbird these sounds are produced by the syrinx (Suthers and Hector 1985). Under controlled laboratory conditions Kellogg (1962) showed that blind humans can echolocate the relative distance and size of wooden discs placed in front of their faces and can also discriminate between discs of different texture (e.g. cloth versus wood). One observer proved capable of detecting a change in the position of a 30 cm diameter disc placed 60 cm away from him when that disc was moved 10 cm nearer or farther away. As an indication of performance in echolocatory size perception, one blind observer could reliably discriminate between two wooden discs of 24 and 26 cm diameter placed 30 cm in front of his face.

LIMITATIONS UPON ECHOLOCATION IN BIRDS

The relatively poor echolocatory performance of the Oilbird and swiftlets (and blind humans) compared with bats and dolphins can be explained by reference to the frequencies of sound which are used. The physics of echolocation, although complex, is well understood [see Pye (1979) for a brief review], and it is clear that fine spatial discriminations require high frequency sounds, outside the range of avian and human hearing, known as ultrasound.

Bats, rodents and cetaceans are among the many groups of mammals which are capable of producing and hearing ultrasounds. In many mammalian species hearing extends beyond 100 kHz and in some bats may extend to 160 kHz (1 kHz = 1,000 cycles per second). Birds and humans on the other hand are restricted in their hearing to relatively low frequency sounds. In the case of humans the normal ear is most sensitive to sounds at about 2 kHz (about 3 octaves above "middle C" on the piano) and sensitivity declines steadily with higher frequencies until sounds above 20 kHz are inaudible even in young people (high frequency hearing is progressively lost with age).

In birds the upper limit to hearing is even lower than in humans. No birds have yet been recorded with sensitivity above about 12 kHz; greatest sensitivity in all birds so far investigated lies in the range 1–5 kHz (Dooling 1982). [It is noteworthy that what is commonly referred to as the "high pitched" songs of the smaller passerines, rarely contain sound frequencies above 8 kHz with most sound energy at or below 4 kHz (Catchpole 1979). The highest note on most pianos produces a sound frequency of just over 4 kHz.] In the Oilbird, greatest sensitivity to sound occurs, as in man, at around 2 kHz, beyond which it declines rapidly and the bird appears to have little or no sensitivity to sounds above 6 kHz (Konishi and Knudsen 1979). Konishi and Knudsen demonstrated that most of the energy in the brief sound pulses (clicks) produced by the Oilbird as it echolocates is in the region of 1.5–2.5 kHz and therefore matches the frequency range to which the Oilbird's hearing is most sensitive. The use of such low frequency sounds for echolocation thus limits the degree of spatial detail that can be resolved by the Oilbird.

Comparable information on the hearing of the swiftlets is not available although echolocatory calls of the Uniform Swiftlet *Aerodramus* (formerly *Collacalia*) *vanikorensis* is reported to contain most energy in the 4.5–7.5 kHz region (Griffin and Suthers 1970), i.e. slightly above that of the Oilbird but still very much lower than in the bats and not within a frequency range that can theoretically yield fine spatial detail.

It would seem that the relatively poor echolocatory performance of the Oilbird and swiftlets is due to their inability to use other than relatively low frequency sounds. It might be expected that natural selection would have favoured the evolution in these birds of an echolocatory mechanism employing sounds of much higher frequencies, and more directional mechanisms of hearing comparable to those used by bats. This would presumably have had particular utility in the case of the swiftlets which, like many bats, not only roost inside caves but also feed upon aerial insects. However, evidence on the structure and evolution of the avian ear (Masterton *et al* 1969; Manley 1972) suggests that such evolution may not have been possible due to constraints on the development of the avian ear, especially the cochlea.

The cochlea is the part of the vertebrate inner ear in which sound frequencies are first analysed by the auditory system. In general, longer cochleas are correlated with the perception of higher frequencies, and in mammals the cochlea is coiled upon itself like a clock spring, thus permitting greater length relative to the small volume of the skull (Manley 1972). In birds, however, the cochlea is only ever straight or slightly curved and consequently short compared with the coiled cochleas of mammals (Smith 1981, 1985). Thus it seems that because of their particular anatomy birds cannot achieve high frequency hearing. It is unlikely therefore that any bird could echolocate with an accuracy comparable to that of the bats.

BIRD FLIGHT GUIDED BY ECHOLOCATION

It can be seen that the cave dwelling birds present a particularly interesting set of solutions to the problems of bird flight in total darkness. The following general points seem noteworthy. First, it should not be overlooked that these birds are prepared to fly without the benefit of any visual guidance, and so it cannot be concluded as an invariable rule that visual guidance is an essential prerequisite for flight in birds. However, it should be noted that the Oilbird is capable of very slow flight and is even able to hover momentarily. Secondly, although they have the benefit of echolocation to guide them, the spatial detail which they are able to detect by this means is relatively gross. Thus Oilbirds are capable and willing to undertake flight even though fine spatial detail cannot be perceived by any sensory means. Thirdly, flight guided by auditory cues takes place only within the spatially simple interior of a cave which is not only a stable environment, but also one that is likely to be well known to each bird.

The Oilbird, for example, appears to form a long lasting monogamous pair-bond which is centred on the nest site, where, if undisturbed by human interference, both individuals roost throughout the year. Snow (1962) suggests that individual birds may be attached to the same nest site throughout their lives (life expectancy is unkown but it is likely that the birds are relatively long lived). Thus, for the Oilbird, life within its cave is highly predictable, certainly from day-to-day and possibly from year-to-year. Whether the same is true for the swiftlets is less certain, but Langham (1980) noted that individual Edible-nest Swiftlets were faithful to their roosting sites within a colony over a seven-month period.

OLFACTION AND CAVE DWELLING

Although the use of active sonar and a predictable life style within a spatially simple environment are important components in accounting for the Oilbird's behaviour in total darkness, it is likely that other sensory cues are also important. For example, Snow (1961) has remarked that tactile cues, especially those perceived via the group of long rictal bristles around the margin of the upper mandible, may be important in communication between birds (adults and young) on their nesting or roosting ledge. Also noted by Snow was the possibly important role of olfaction both in locating individual nest sites and individual birds (Oilbirds apparently have a characteristic odour). Although neither of these sensory abilities have been the subject of experimental investigation, anatomical studies do support the idea that olfaction may be of particular importance in the Oilbird.

Bang and Wenzel (1985) in reviewing anatomical studies of avian olfactory apparatus noted that, "The Oilbird and Turkey Vulture *Cathartes aura* have roughly the same total amount of olfactory equipment" (p. 206). The importance of this comparison is that the olfactory apparatus (both the receptive surfaces inside the nasal cavity and the areas of brain devoted to olfaction) in the Turkey Vulture (a New World Cathartid vulture) is amongst the largest so far described in birds (Bang and Cobb 1968). Furthermore, this species may be able to locate food from a considerable distance solely by smell [cf. some of the Procellariformes discussed above which can locate food items

Foraging Oilbirds *Steatornis caripensis*, may hover for brief periods.

in this way, also have olfactory apparatus of comparable size to that of the Oilbird (Hutchison and Wenzel 1980)].

Snow (1961) suggested that not only could the Oilbird use smell to assist in the location of its mate or young within the cave, but also that smell could be used to locate trees bearing ripe fruit (the exclusive diet of the Oilbird) and even to locate individual fruits. Using radio tracking, Tannenbaum and Wrege (1984) showed that Oilbirds may forage up to 8 km from their cave in search of specific food sources. Many of the fruits sought by the birds occur in large bunches near the tops of trees and so they could provide a significant olfactory beacon for birds flying close to the tree canopy. It is also possible that such bunches could be detected in silhouette against the sky, at least under certain night-time conditions. Unfortunately little is known about the visual capacities of Oilbirds. Their eyes are relatively large and it has been established that they are certainly functional. Pettigrew and Konishi (1984) have conjectured that their vision may be similar to that of Barn Owls *Tyto alba*.

The possible role of olfaction in the behaviour of the swiftlets has not been investigated, although Bang and Wenzel (1985) do suggest that smell may be important in these species since they found that in the Glossy or White-bellied Swiftlet *Collocalia esculenta* the olfactory apparatus does seem well developed with a rich olfactory epithelium (the surface within the nasal cavity where olfactory stimuli are detected).

SENSORY AND COGNITIVE ASPECTS OF CAVE DWELLING

The above discussion has presented a mixture of information on the sensory capacities available to cave-dwelling birds and on their natural history. With this information it is possible to account, at least in part, for their mobility within the caves. It has been shown that the spatial information available to them when in the cave is rather coarse, and neither echolocatory nor olfactory cues can reveal fine detail. It may be proposed therefore that their ability to fly, albeit rather slowly, within a totally dark interior of a cave requires an integration of what little spatial information that is immediately available through the senses, and a knowledge of the spatial structure of the cave. Thus it seems that knowledge of the layout of the cave is used to usefully intrepret the rather coarse echolocated details which are received during flight.

There are anecdotal observations which suggest that the swiftlets may be so familiar with the structure of their caves that they choose to ignore, or do not constantly gather, echolocatory information. Smythe and Roberts (1983) report that Grey Swiftlets, "are initially unable to detect the presence of even large objects, such as a human body, when placed in a normal cave flyway" (p. 344). This suggests that these birds are perhaps not bothering to send out echolocatory calls to sample the structure of their surroundings as they fly in the total darkness of the familiar cave.

In order to enjoy the advantages of a cave as a safe roosting and breeding site, however, it would seem that individual birds must remain faithful to a particular cave, and even to a particular ledge within that cave, for a long period of time, possibly throughout the bird's whole life. Such a restricted life style would seem to apply to the Oilbird, but it is not known if it applies as strictly to swiftlets. Certainly the problem of locating an individual nest within a colony which may contain hundreds (in the case of the Oilbird) and sometimes over a million individuals (in the case of the swiftlets) closely packed together, does suggest that spatial memory must be very important, although passive sonar, based upon the calls of mates, and olfactory guidance to individuals may play a part in the final stages of reaching a nest or roosting ledge.

In the case of swiftlets, mobility outside the cave does not need to be addressed here since they are diurnal and feed much in the same way as other swifts, by trawling for prey often through very dense aggregations of insects. However, the Oilbird is nocturnal and feeds on fruit, and seems to be the only bird which can be so described. How this is achieved is not clear, but Snow (1961) suggests that to some extent the birds have prior knowledge of the position of suitable foraging trees and that they are able to find them using olfactory cues. It should be noted that outside the cave the birds rarely echolocate. Furthermore they are thought to fly usually above the tree canopy, in the open airspace, and to pluck fruit whilst in hovering flight rather than when flying among the branches. The degree to which bunches or individual items of fruit are located by smell or sight would clearly be of interest.

CHAPTER 7

Regularly nocturnal birds

With the exception of the Oilbird, all of the species discussed hitherto have been examples of birds which are only occasionally active at nocturnal light levels. It was seen that while such behaviour is relatively common, these birds are active outside daylight hours only under specific, often well-defined, circumstances. Also those activities which are conducted at night are usually relatively restricted. Certainly these birds do not at such times simply go about their usual day-time routine at night. Thus the occurrence of such occasional nocturnal or crepuscular activity usually applies to specific aspects of a bird's life cycle. These include, for example, the night-time foraging of waterfowl and waders, nocturnal visits to nest burrows in shearwaters, or nocturnal migration along certain sections of a route in passerines.

Although these occasional nocturnal activities encompass very different behaviours, they do share a number of important common themes.

First, there is evidence that such nocturnal activities are often executed less efficiently than the same behaviour during day-time. There is evidence that wading birds may forage less efficiently at night than during the day, even on exactly the same foraging area. Petrels and shearwaters returning to their nest burrows at night often stumble into obstacles or may choose the wrong nest burrow; and nocturnally migrating birds sometimes die as a result of crashing into illuminated obstacles, such as office blocks or lighthouses, to which they are attracted. There is no evidence that birds migrating by day ever do so.

Secondly, when occasional nocturnal activities involve flight, they take place in open habitats well away from obstacles and certainly clear of complex vegetation. The foraging of waders and waterfowl is done in wide open habitats, and nocturnal bird migration takes place in the upper airspace. Similarly, the twilight foraging of skimmers and falcons occurs in open areas free of hazards. In those instances where nocturnal activity does occur in complex habitats they seem rarely to involve flight. The nocturnal singing of the Nightingale is from the same few perches each night, while other nocturnally vocal species, such as the Water Rail and Corncrake, walk on the ground among relatively dense but uniform vegetation.

Thirdly, where occasional nocturnal activity involves foraging, there is evidence that the actual location and selection of food items is principally achieved by senses other than vision; indeed even the daylight foraging techniques of these species tend not to rely upon visual cues. For example, tactile and taste sensitivity in the bill play major roles in the foraging of many night-feeding waterfowl and waders. In a similar way, tactile sensitivity in the bill would seem to be crucial in the foraging of skimmers, and olfactory cues of prime importance in the nocturnal foraging of petrels. Thus, in one sense, the sensory cues employed by these birds when foraging can be seen as predisposing them to foraging at night, compared with the visually guided day-time foraging of most birds. Certainly, the limited experimental data on how foraging efficiency in small passerines is influenced by light levels, suggests that these birds are unable to forage with high efficiency until illumination exceeds that experienced at sunrise.

Much of what has been stated above would support the idea that many of these occasionally nocturnally-active species are not sensorily well equipped to cope with all the sensory problems which the nocturnal environment actually presents to a flying bird. It is probably for this reason that the actual behaviours executed at night are so restricted.

It can also be concluded that, for many species, nocturnal activity is a less-preferred behavioural strategy than activity during day-time. This is perhaps exemplified most clearly in the nocturnal migratory behaviour of passerines and the foraging of waders. In the case of the night migratory behaviour, there is evidence that at least some of these species migrate at night only along certain sections of their route. Furthermore, there would seem to be good evidence that when flying at night passerine migrants may become confused by visual cues from the ground, the product of which is the incidents of birds being attracted to lighthouses and other artificial light sources under particular conditions.

That the night foraging of waders is not a preferred strategy is suggested by the following evidence. Although such behaviour occurs primarily because the tidal cycle may prohibit foraging during daylight hours at certain times of the lunar cycle, waders by no means always forage at night when feeding areas are exposed at this time. They are more likely to forage at night when food requirements are particularly high, often during harsh winter weather or prior to migration. At other times the birds feed during those daylight hours which are available, and roost during darkness. They do not seem to feed by night whenever the opportunity arises.

It is against this background that attention now turns to birds which are habitually nocturnal. That is, bird species which are active at night throughout the year and complete all aspects of their life cycle at this time. Only occasionally are these birds active by day. Given the restricted nature of the nocturnal activities discussed above and their particular sensory problems, the question now arises as to how it is that these truly nocturnal birds are able to complete all aspects of their life cycle in the hours of darkness. The remainder of this book is devoted principally to seeking an answer to this question.

NOCTURNAL, FLIGHTLESS BIRDS

As discussed above, the principal restrictions upon occasional nocturnal activity in birds seem to arise from problems associated with, (i) the control of flight in structurally complex habitats and (ii) the use of vision in foraging. It is perhaps not surprising therefore to find that some of the terrestrial flightless birds of the world are completely nocturnal. However, even for these flightless birds a nocturnal life style would seem somewhat restrictive, and by no means are all flightless birds nocturnal in their habits. Indeed among the flightless birds of the order Struthioniformes, a nocturnal habit has only evolved in the three smallest species, the kiwis. It has been suggested by Fedducia (1985) that these birds have "developed nocturnal habits instead of relying on fleetness of foot for safety". All the extant larger species of this order (Ostrich, Rheas, Cassowaries and Emu) are capable of running very fast and of inflicting serious injury to any animal which may attack them (including man). However, Reid and Williams (1975) have pointed out that prior to the advent of man on New Zealand some 1000 years ago, kiwis were probably largely free from predation, which suggests that foraging, rather than escaping from predation, may be a more important consideration for the evolution of nocturnality in these birds.

The three species of kiwi, all of which are endemic to New Zealand, are the most well-known species of truly nocturnal flightless birds. However, there are a few other birds which are not truly flightless but apparently rarely fly (or are only capable of un-powered gliding flight to the ground from trees), which are also active at night, these include three species of parrot: the Ground Parrot *Pezoporus wallicus* and the Night Parrot *Geopsittacus occidentalis* of Australia, and the Kakapo *Strigops habroptilus* of New Zealand (Forshaw 1978). It was at one time thought that the Kagu *Rhynochetos jubatus* (a bird from New Caledonia whose taxonomic position is uncertain but which is now included in the Gruiformes) was both flightless and nocturnal. However, this bird is very shy and secretive and detailed studies of captive birds have now shown that it is capable of flight (though it rarely does so) and is primarily diurnal in its activities (Walters 1978). It seems that the Kagu is only very vocal at night during the breeding season and may be regarded as nocturnally active during the period of incubation only (Neyrolles 1985). Thus the Kagu would seem to fit more appropriately into the category of an occasionally nocturnal species.

THE KIWIS

The three species of kiwis have roused great interest ever since they first became known to modern ornithology at various dates during the last century. Their taxonomic status is still not fully agreed. Some authors, following Peters (1931) *Check-list of Birds of the World,* have placed the kiwi family, Apterygidae, in an order of their own, the Apterygiformes (Reid and Williams 1975; Dawson 1978). However, the recent revisions by Voous (1985) and Sibley *et al* (1988) based upon biochemical classification techniques place the family in a Suborder (Apteryges or Casuarii), both within the Struthioniformes.

Whatever their status, all authorities agree that these completely flightless birds are, perhaps with the exception of birds from Stewart Island, totally nocturnal. Considerable effort has been expended on understanding their life style in their preferred habitat (the floor of the New Zealand wet forests) and in captive collections. The sensory basis of their foraging has also received some attention. A comprehensive account of their general biology is given by Reid and Williams (1975).

Field observations together with experimental and anatomical studies all attest to the importance of olfaction in the kiwis. Anatomically, the Brown Kiwi *Apteryx australis* exhibits three important features which demonstrate the importance of olfaction in guiding its behaviour. First, the proportion of the brain devoted to olfaction (the olfactory lobes) is among the largest of all birds, and by far the largest of any terrestrial bird species (Bang 1971). The only other birds which have such well developed olfactory lobes are all sea birds from the order Procellariiformes. Secondly, the elaborations of the surfaces where olfactory stimuli are detected inside the nasal cavity are extremely complex. This gives the bird a large surface area upon which olfactory stimuli can be received. Thirdly, the kiwis are unique in being the only birds with nostrils that open at the tip of the bill (bill length approx. 11–15 cm) rather than near the base (Bang and Wenzel 1985). Reid and Williams (1975) describe how, when foraging, kiwis can be heard by virtue of snuffling noises which are "made by expelling air forcibly through the nostrils while feeding. This is clearly audible at many metres distance as the bird wanders about in the forest at night, probing continuously with its bill" (p. 303).

Experiments on the ability of kiwis (species not specified) to locate earthworms buried beneath the soil at the depth of about 25 mm showed that they could locate their prey using olfactory cues alone (Wenzel 1968). However, although kiwis are noted for taking invertebrates by digging with

All three species of Kiwi *Apteryx* spp. have a highly developed sense of smell useful for locating buried prey.

their bills, they are regarded as omnivorous, also eating fruits, seeds and fibrous plant material. The extent to which olfaction plays a part in the detection of these latter food items is unknown. It might be expected that touch sensitivity in the bill tip (similar to that found in the long-billed wading birds, Chapter 4) would also play a part but there is no evidence of this.

There is also scant data on the use of vision and hearing in kiwis. Their eyes have been said to be "as degenerate as a vertebrate eye can be" (Walls 1942) and that they "perhaps have vision akin to that of moles" (Reid and Williams 1975, p. 316) but what this might mean functionally is unclear. By avian standards the eyes of kiwis are small for a bird of this size, being only 8 mm in diameter (Kajikawa 1923), and this may have led to the assumptions that the eyes are non-functional or degenerate. Kiwi eyes are, in fact, as large as those of many visually guided passerine species, such as the European Starling *Sturnus vulgaris* (Martin 1986c) and larger than those of the rat, a nocturnally active mammal (Hughes 1979) known to be able to carry out reasonably complex visual discriminations. A more recent study of the refractive state of the eyes of the Brown Kiwi *Apteryx australis* has shown that they are well focused (Sivak and Howland 1987). This suggests that kiwi eyes are functional and not degenerate, though the actual visual capacities remain unknown.

Reid and Williams (1975) suggest that hearing "is probably important and fairly well-developed" in kiwis, but these observations are based solely upon anecdotal observations rather than critical studies.

The picture which emerges from these field observations and somewhat limited experimental data is that these flightless birds tend not to rely on vision to guide their behaviour beneath a tree canopy at night. Certainly, olfactory cues are employed to locate food items within the soil, although the possible role of hearing (leaf litter rustle produced by invertebrates) and tactile cues (kiwis have long bristles extending from around the base of the bill) cannot be overlooked. It is possible that olfaction, hearing and tactile cues may mediate the social behaviour of kiwis in much the same way as in nocturnal mammals which live on the forest floor in other parts of the world. However, there is no evidence that kiwis use scent marking to regulate their social behaviour, as is commonly found in the mammals (Albone 1984).

THE NOCTURNAL PARROTS

The biology, especially the senses, of the three species of mainly nocturnal, near-flightless, parrots are less well known than that of the kiwis. Like the kiwis, the Kakapo, Night Parrot and Ground Parrot live amongst dense ground vegetation (heath, swamp and tall tussocky grasses), and in the case of the Kakapo may also be found under a woodland canopy, in relatively remote regions of New Zealand (Forshaw 1978). The Kakapo is considered to be one of the world's rarest and most endangered species of vertebrate (Merton *et al* 1984), there being less than 50 individuals in existence. Accurate information does not exist, as both sexes lead solitary lives in particularly difficult and inaccessible habitats. The first photographs of young birds at a nest were not published until 1984, when the suspected lek behaviour of the birds was also confirmed (Merton *et al* 1984). A general description of the natural history and efforts to conserve the Kakapo is given in Cemmick and Veitch (1988).

Even less is known of the habits of the Night Parrots. Forshaw (1978) describes it as "one of Australia's most mysterious birds" (p. 276), there having been "no authentic records for more than fifty years". However, reports of various sightings have led Forshaw to conclude that the species does still exist. During the day they are reputed to shelter in tunnels in the base of grass tussocks and even conceal the entrance with pieces of grass. There are reports that the Night Parrot can fly well but, apparently, it rarely does so. The Ground Parrot is better known than the other two species, but even so much of its biology has not been described. They apparently favour coastal habitats (salt marsh and swamps), are said to have a weak flight and are believed to be sedentary.

The Kakapo is considered to differ sufficiently from all other parrots to be placed in a sub-family of its own (Strigopinae). Like the other two parrots the Kakapo is not completely flightless; it is in fact only incapable of powered flight. The bird can climb trees using its claws and bill and will on occasion glide down to the ground. However, for all intents and purposes it is a ground living, nocturnal bird.

All three species are herbivorous, eating a wide range of leaves, roots, fruit and seeds and thus, unlike the kiwis, do not have to detect and capture mobile prey. In the case of the Kakapo, it seems that the birds live a very regular existence, visiting known food sources for which they may use a system of regular paths (Shepard and Spitzer 1985).

Unfortunately nothing is known about the sensory bases of the foraging behaviour of these birds. However, it is known that parrots have a well developed bill tip organ (Gottschaldt 1985) [the first description of the avian bill tip organ was in a parrot (Goujon 1869)] and the highest density of taste receptors reported in any bird so far investigated (Berkhoudt 1985). Both touch and taste sensitivity probably play an important role in the selection of

The nocturnal flightless parrot, the Kakapo *Strigops habroptilus*.

food items, but it seems unlikely that such sensitivity is responsible for their initial detection. There have been no systematic studies of the sense of smell or of the olfactory apparatus in any species of parrot. The Kakapo is, however, noted for having a characteristic odour which is detectable by man, and so it seems possible that olfaction could also play a part in the social behaviour of these birds.

It would seem reasonable to assume that the ancestors of all three species of nocturnal parrot were diurnal and capable of powered flight. It is interesting to note, therefore, that a nocturnal habit and flightlessness appear to be the result of ecological convergence in these species. The evolutionary path by which this came about is not clear, though it seems likely that the birds' herbivorous diet, perhaps relying upon tactile and taste cues within the mouth for the selection of food items, may have been an important element in the evolution of nocturnal foraging. It is impossible to say whether a terrestrial life style predisposed the birds to a nocturnal habit, or whether, conversely, adopting a nocturnal life style led to the birds becoming increasingly terrestrial because of the problems of flight in complex habitats at night. Whatever the evolutionary route, it is noteworthy that today the only three species of parrot (out of about 330 extant species) which are terrestrial are also the only ones which are nocturnal in their habits.

NOCTURNAL FLYING BIRDS

Under this heading shall be considered birds from just three orders: the Caprimulgiformes (nightjars and their allies), the Strigiformes (owls and Barn Owls) and the Charadriiformes (wading birds and their allies). Within these three groups are birds which are capable of flight and are likely to be active on most nights of the year. Whether it can be definitely concluded that all of these species complete all aspects of their life cycle exclusively between the hours of dusk and dawn is not clear. What is known, however, is that all of these species are regularly active at night throughout the breeding season and that they also forage at night.

Although all birds of the three orders are capable of sustained flight it will be seen that one group, the nocturnal wading birds, is, however, essentially terrestrial and feeds exclusively on the ground. It must also be noted from the outset that in none of these orders are all species nocturnal. Indeed, even among the owls, it seems likely that it is only a minority of species which is exclusively nocturnal. Some owl species may be more or less strictly diurnal; others are crepuscular but may also hunt by day and by night. Among the Caprimulgiformes similar considerations may also apply. It has already been noted above (Chapter 4) that among the Charadriiformes occasional nocturnal activity may be recorded. The Charadriiform species included here as regularly nocturnal are a small minority (13 species) of the total in this order. However, even among these birds nocturnal foraging may not occur throughout the year.

It should also be noted that data on the extent of nocturnality in many of the species discussed are based upon anecdotal and casual field observations, rather than systematic studies. As was seen in Chapter 2, light levels at night can vary markedly, and it may be that some reports of night-time activity actually refer to nights during which light levels remained relatively high due to bright moonlight and/or a combination of latitude and season. As was explained in Chapter 4, a bird may be recorded as active at night but that does not necessarily make it a nocturnal species within the definition introduced in Chapter 3 (a strictly nocturnal bird is one which habitually completes all aspects of its life cycle between the hours of sunset and sunrise, or when ambient light levels are below those produced in open habitats at sunset or sunrise). This consideration applies to the records of nocturnal activity in owls and nightjars as much as to species of other orders.

NOCTURNALITY AMONG THE CHARADRIIFORMES

In Chapter 4 it was pointed out that some species of the Charadriiformes occasionally foraged at night-time. In the main those birds were species which foraged by probing into soft mud or soil. It was argued that they were, at times, able to forage at night because they frequented open habitats devoid of obstructions. Also, that their foraging technique depended principally upon "blind probing" in which prey detection and selection is achieved principally by tactile and taste sensitivity in the bill. However, in the case of the plovers

(Charadriidae) auditory and visual cues may also serve to attract birds to probe in a specific area. There is, however, a further group of species from this order which is more regularly nocturnal. The group shares with the occasionally nocturnal birds a preference for open habitats with sparse and/or short vegetation. They tend not to probe but take prey directly from the surface and, as in the plovers, their night-time foraging may be guided by sound cues as well as by vision.

Regular nocturnal activity is restricted among the waders to 13 closely related species from three genera: the coursers of the genus *Rhinoptilus* (Glareolidae), and the stone-curlews and thick-knees of the *Burhinus* and *Esacus* genera (Burhinidae). All of these species, with the exception of the Stone Curlew *B. oedicnemus*, are found more or less exclusively within tropical latitudes. All of them prefer habitats which are open, flat or rolling, with sparse or very short vegetation. Some species, such as those of the genus *Esacus* (the Great Thick-knee, *E. recurvirostris*, and the Beach Thick-knee *E. magnirostris*), show a preference for coastal breeding and feeding sites, but other species prefer open, dry, inland situations, though often with water in the vicinity. Even in Violet-tipped Courser *R. chalcopterus*, which may be associated with dense scrub and open woodlands, there is a preference for open areas where bare earth is exposed. Perhaps the overriding characteristic of all of these species' preferred habitats is all-round visibility at ground level with no impediments to running [Haymen *et al* (1986) summarise the habitat and food preferences for all of these species].

All of the above species are primarily sedentary, except for the more northerly breeding populations of Stone Curlew *B. oedicnemus* in Europe and Asia, which migrate to Africa and India in the non-breeding season, and the Violet-tipped Courser which is thought to migrate within southern Africa.

Hearing may help the Stone Curlew *Burhinus oedicnemus* and other species of thick-knees to locate large insects at night.

When feeding, these birds rarely fly, preferring instead to run across the open terrain in pursuit of prey. Even if approached by humans they tend to stay put, relying upon camouflage, rather than take to the wing. All of them feed mainly upon large terrestrial insects and other invertebrates, but will also take small mammals, lizards and crabs when available. All such prey is usually taken from the surface rather than by probing or in flight.

As in the case of the nocturnal parrots, nothing is known of the sensory capacities of these birds. However, they do have large eyes, which suggests that vision may have an important functional role in the guidance of nocturnal behaviour. However, it is likely that nocturnal foraging could be accounted for by the use of hearing as well as vision. Their prey are relatively large and are likely to be heard as they move across dry terrain, or they may emit sounds themselves. For example, insects such as crickets and grasshoppers are important in the diets of many of these birds. Such invertebrates may be particularly active after dusk, calling or singing constantly. Indeed, Hald-Mortensen (1970) noted that, on Tenerife, Stone Curlews started foraging when natural illumination levels fell to between 2 and 20 lux (0.3–1.3 log), which is in the lower range of civil twilight (Figure 2.5). Although this suggests that these low light levels may be important in controlling the foraging of these birds, light levels may not be acting directly. Hald-Mortensen noted that such light levels were exactly those at which crickets (*Gryllus* spp.) usually began to sing. Field observations of the foraging of Stone Curlews in England by Scott (1965, reported in Cramp and Simmons 1983), also support the idea that hearing may play an important part in the detection of their insect prey.

Thus, in the occasionally nocturnal plover species (Charadriidae) and in these more regularly nocturnal Stone Curlews and thick-knees, hearing may play an important role in the detection of prey. In the case of the plovers, the birds may detect the approximate location of prey by hearing its spasmodic movements and then seizing it by probing in the mud or soil, guided by tactile cues (Fallet 1962). The Stone Curlews and coursers may be guided to prey entirely by sound cues made irregularly or constantly by the prey. In Stone Curlews and thick-knees, visual cues may also be of some assistance in the final approach to the prey but these are of less value to the plovers since their prey is often buried in mud or hidden beneath vegetation. In both types of foraging, approaching the prey on foot in structurally simple environments [Hayman *et al* (1986) remark that some species of thick-knee are particularly noted for foraging on roadways] could perhaps be achieved with the aid of only minimal visual cues.

NOCTURNALITY AMONG THE CAPRIMULGIFORMES

The order Caprimulgiformes contains about 100 species divided between five families (Voous 1985). All species are nocturnal and/or crepuscular to varying degrees, but within these five families there are important differences in habitat preferences and foraging techniques.

The unique biology and life style of the Oilbird has been discussed above (Chapter 6). It is the only bird species which is both nocturnal and feeds on a diet of fruit. However, the means by which it detects the ripe fruit is unknown but there is some indirect evidence that olfaction and vision may be important. There is no evidence that olfaction plays a role in the foraging of any other Caprimulgiform species.

Woodland species: frogmouths, potoos and owlet-nightjars

Flight under nocturnal light levels in the open above the woodland canopy may not perhaps present particularly exacting problems for the control of flight in the Oilbird. However, three families of birds (containing 26 species) in the Caprimulgiformes are regarded as strictly nocturnal, but because they live within or beneath a woodland canopy it seems reasonable to assume that they must fly at night within a more spatially complex environment than any of the species so far mentioned. It will be seen, however, that their foraging techniques all have features in common which can account, at least in part, for the birds' ability to forage at night within a woodland. However, flight under the woodland canopy cannot be so readily accounted for.

The three families of nocturnal/woodland species of the Caprimulgiformes are the frogmouths (Podargidae), potoos (Nyctibiidae) and the owlet-nightjars (or owlet-frogmouths) (Aegothelidae). Details of their biology and natural history can be found in Fleay (1968), Skutch (1970), Harrison (1978), Schodde and Mason (1980), and Serventy (1936, 1985a,b). All of the species live within the tropical and sub-tropical climatic zones. The frogmouths are distributed around southeast Asia, Indonesia and Australia; the potoos are found only in the American tropics, and the owlet-nightjars are from New Guinea and Australia. All the species are more or less sedentary and solitary in their habits.

All of the above feed primarily upon large insects and other invertebrates, although the larger species of frogmouths (such as the Tawny Frogmouth *Podargus strigoides*) also take small vertebrates such as frogs and mice. One striking feature which all frogmouths have in common is their large gape. In some species the bill is surrounded by rictal bristles on the edge of the upper mandible. Although the large gape would suggest that it is well suited for trawling insects out of the air, this foraging technique is not used by any of the species. The frogmouths feed principally by taking prey from the ground after pouncing from a low perch or running along the ground, the bird actually striking or capturing the prey directly with its rather heavy bill (Serventy 1936). The owlet-frogmouths also take prey from the ground but they tend not to use an elevated perch for the approach, preferring to pursue prey along the ground. Again, the prey is taken directly with the bill. They have been recorded as occasionally snatching insects from the air but they seldom pursue them in flight and, on the whole, the diet consists of non-flying invertebrates such as millipedes, spiders and ants. The potoos do, however, take insects from the air but, rather than trawl them, these are taken "flycatcher fashion"

The large gape of the Tawny Frogmouth *Podargus strigoides* is used to take relatively large prey directly in a pounce from the ground or from a low perch.

by sorties from a fixed perch to which the bird returns. Potoos generally live in more open woodlands (sometimes cultivated lands and plantations) than the frogmouths, which prefer more densely forested areas.

Nothing is known about the sensory capacities of these birds although, as in the nocturnal waders discussed above, the eyes are relatively large, thus suggesting that vision is of importance to their nocturnal activities. The influence of light levels on their foraging has not been investigated.

The possible role of hearing as a means of locating prey, especially on the ground, may be particularly important in the prey capture techniques of the frogmouths and the owlet-frogmouths. The latter species in particular could perhaps be seen, with regard to foraging, as a woodland equivalent to some of the nocturnal Charadriiformes. Both groups forage primarily on and from the ground for large invertebrates which are approached on foot.

The rictal bristles found around the edge of the gape and the modified filoplume feathers on other parts of the head in many of these species, are often assumed to aid in the detection and capture of prey. This can be achieved either by the bristles physically guiding insects towards the open mouth, and/ or by providing tactile cues to guide the bird's mouth towards a prey item that has touched the bristles. The base of most feather follicles are particularly rich in Herbst corpuscles (Ostmann *et al* 1963), and so any feather or bristle could act as an extended mechano-receptor. Unfortunately there appear to be no experimental studies of the role of rictal bristles in prey capture in any species of birds.

The picture which emerges of these nocturnal woodland birds is of species similar in a number of respects to the nocturnal wading birds but also showing some similarities with the woodland owls (Chapter 8). First, although these Caprimulgiforms clearly can fly, their prey capture techniques either do not necessitate flight or, if flight does occur, it is of short range from fixed perches down to prey and back again. Secondly, it is possible that hearing plays a role in the location of prey, although there is no experimental evidence for this or data on their abilities to locate sounds. The potoos, which generally catch aerial insects from a fixed perch sortie, do seem to need to be visually guided towards individual prey items. Thirdly, many of these species tend to be both solitary and sedentary at all times of the year.

Open habitat species—the nightjars

It has been pointed out a number of times that much nocturnal and occasional nocturnal behaviour in birds requires an unobstructed habitat. This may vary from the open airspace in which nocturnal migrants fly, the open conditions of estuarine mud banks, or the barren terrain of a semi-desert. However, none of these species actually forage in the open airspace. The nocturnal and/or crepuscular species of nightjars of the family Caprimulgidae also employ open habitat, but they forage regularly in the open airspace where they use specialised techniques to take flying insect prey.

There are approximately seventy species of nightjar in the Caprimulgidae more than half of which fall within a single genus, *Caprimulgus*, the best known species being the Nightjar *C. europaeus* which occurs throughout much of Europe and Asia. Its biology is described in detail in Cramp (1985). All species of nightjar tend to spend the day roosting on the ground, or sometimes in trees, and become active around dusk when they begin foraging for insects, which are the exclusive diet of these birds. They tend to take the larger flying insects, especially moths and beetles, but smaller insects and flightless forms are not uncommon in their diet. The habitats preferred both for breeding and foraging include heathland, broad woodland rides, grasslands, tree plantations in their early stages and burnt areas (Berry 1979; Ravenscroft 1989). For some species such as the Common Nighthawk *Chordeiles minor* and the Egyptian Nightjar *Caprimulgus aegyptius*, even the airspace above towns (where they may breed on flat roof tops) are used. Some species may also feed over lakes and rivers. The one common feature is that none of them appears to forage amongst vegetation. Where trees are present, as for example along woodland rides and in regenerating heathlands and plantations, the birds forage above the vegetation, not within it. Trees around the periphery of open sites may be important as song posts.

The actual foraging techniques, and the specialised anatomy of the mouth parts associated with foraging, have been the subject of a number of investigations. Data on the foraging techniques even for a well-studied species (e.g. *C. europaeus*) are, however, somewhat contradictory. For example, it is not clear whether nightjars ever capture prey by "open billed trawling" through dense clouds of insects, as against the pursuit of individual prey items (see

The large gape of the European Nightjar *Caprimulgus europaeus* is often used to trawl through clouds of small insects as they emerge from vegetation into the open airspace at dusk.

discussion in Cramp 1985). Also, there are conflicting opinions as to whether they take prey when on the ground or only in flight. It is not clear to what extent the actual light levels of twilight and night-time affect the particular foraging techniques employed. Certainly it is possible that the differing observations on foraging may relate to different light levels as well as to different prey availabilities. Analysis of the diet, however, shows that even the Nightjar will take insect prey of a very wide size range, from mosquitoes and micro-moths (Microlepidoptera) to large moths, beetles and cockroaches. While these larger prey may be taken by individual pursuit, it is difficult to believe that the smaller prey are captured in this way. Indeed the following anatomical evidence on the specialised mouth structure of the nightjars, suggests that they may be well adapted for the capture of insect prey by trawling.

First, all of the species of *Caprimulgus* have rictual bristles growing from around the margin of the upper mandible (these are not present in the nighthawks (Chordeilinae), which are a group of species exclusive to the Americas). The bristles are so arranged that they trap or direct insects towards the open mouth. The sensory properties of the bristles is unknown but it seems unlikely that they would not have some sensory function concerned with feeding (see above).

Secondly, the open gape of these birds is very wide, much wider than might be supposed from inspection of the closed mouth. This is achieved by virtue of a unique structure of the skull (which is quite flexible due to a high degree of pneumatisation and very thin walls), and the arrangement of joints, flexible bones and musculature involving the articulations of the upper and lower mandibles (Buhler 1970, 1981). This unique anatomy spreads the sides of the lower mandible wide apart and results in a nearly circular cross section to the open mouth, as depicted in Figures 7.1 and 7.2.

Thirdly, the roof of the mouth (palate) of nightjars contains a unique sensory structure. In the majority of bird species the palate is lined by a horny sheath which is relatively tough and insensitive to touch, and this presumably functions to protect the palate from damage when eating. In the Caprimulgidae, the horny sheath of the palate is absent and in its place is a highly vascular membrane, bright red in colour because of the number of blood vessels close to the surface. Cowles (1967) who first described this structure concluded that "it would be very sensitive and easily stimulated by an insect striking the surface, enabling the bird to react quickly to the prey while in flight". Cowles was further supported in the idea that this sensitive palate is associated with nocturnal aerial insect-feeding in that, while he was able to show that this structure was present in the 12 species of Caprimulgidae which he examined, it was not present in other Caprimulgiform species, which were either fruit-eating (the Oilbird) or ground feeders (frogmouths and owlet-nightjars).

Figure 7.1   The gape of the Nightjar *Caprimulgus europaeus*. In A and C the jaws are half open and the bones of the lower mandible are spread only slightly apart. In B and D the jaws are fully open and the lower mandible bones spread wide apart. (From Buhler 1970.)

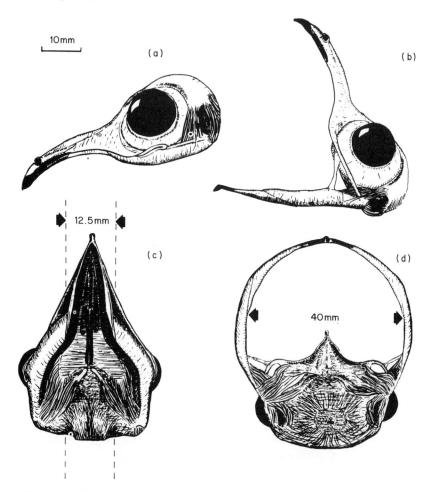

Figure 7.2   The jaw structure and gape of the Red-necked Nightjar *Caprimulgus ruficollis*. The skull is shown from the side (A and B) and from below (C and D). When the jaws are closed (A and C) the width of the lower mandible is 12.5 mm, when it is fully open (B and D) the bones of the lower mandible spread far apart to approximate a circle 40 mm in diameter. (From Buhler 1970.)

Thus it seems that the nightjars have three special features, all of which act to facilitate the capture of insect prey on the wing. The picture emerges of a bird able to trawl with its wide open gape, whose capture area is further enlarged by rictal bristles, and endowed with a touch sensitive mechanism inside the bill enabling even a small insect, when it strikes the palate, to be detected and the mouth snapped shut. It is of course extremely difficult to verify this in the field, since such feeding will by definition take place in the open airspace at night. Field observers are much more likely to witness other types of prey-catching activity, such as sorties from a fixed perch.

The actual light levels at which nightjars become active do not appear to have been measured, although it is recorded that the birds usually begin their activity around the time of sunset, though season, latitude, and presence and phase of the moon, all appeared to modify this from day-to-day (Schlegel 1969, cited in Cramp 1985). Lehtonen (1951, cited in Cramp 1985) recorded that flight activity did not begin until the sun was 5–8° below the horizon (i.e. about the end of civil twilight). This suggests that nightjars become active at about the same light levels (Figure 2.5) reported by Hald-Mortensen (1970) for the Stone Curlews on Tenerife. In the Stone Curlew, it was found that these light levels coincided with the period when its prey species became active. Clearly, it would be interesting to know whether nightjars also become active with the onset of activity in its principal prey species.

Unfortunately nothing precise is known about the visual capabilities of any species of Caprimulgidae. However, there is limited anatomical and field evidence which suggests that vision is at least of some importance in their nocturnal activities. There is also more tenuous evidence that vision is unlikely to provide sufficiently detailed information, of itself, to account for all of the nocturnal foraging activities of these birds.

First, the eyes of some Caprimulgidae are relatively large and are, to date, the only bird eyes known to contain a tapetum in the retina (Nicol and Arnott 1974), although it is not clear whether all Caprimulgiform eyes have this structure. A tapetum is a reflective layer at the back of the eyes. It gives rise to the bright eye-shine of many nocturnally-active mammals such as dogs, cats and cattle. The function of these highly reflective tapeta (which may have different structures in different animals (Rodieck 1973)) is the same in all species: to increase the chance that light will be intercepted by the receptors and thus effect a slight but significant increase in the absolute sensitivity of the eye.

The highly reflecting layer of the tapetum lies behind the layer of photoreceptors in the retina, where incoming light is absorbed and converted into neural signals. Placed behind the receptors in this way, the tapetum reflects to the photoreceptors any light that was not absorbed as it passed through the retina the first time. A tapetum thus serves, roughly, to double the chance that any light which enters the eye will be detected. When we see eye-shine we see light which, having entered the animal's eye, has been reflected from the tapetum towards us.

It should be noted that some eye-shine can be detected in most eyes without a tapetum, if they are viewed in the correct way, and this includes those of humans and other birds. This is often seen, for example, in flash photographs when the flash gun and lens are closely aligned. This form of eye-shine is, in the main, due to a small proportion of the light being reflected before it reaches the photoreceptors at the retinal surface. If a tapetum is present, the eye-shine is typically very bright and easily seen; eye-shine produced without a tapetum is relatively dim.

Throughout the animal kingdom, tapeta are only found in species which are either nocturnally active or live in environments where there is little natural light, for example, deep-sea fish (Rodieck 1973). For animals that live mainly in bright daylight conditions, the increased sensitivity which a tapetum can provide is of no consequence and may be a disadvantage because the light reflected from behind the retina is likely to reduce the eye's ability to resolve fine detail. This will not be disadvantageous to an animal functioning at low light levels because fine detail cannot ever be resolved at such levels (see Chapter 8 below). However, increasing the chances of capturing any of the scarce light that is around at night will clearly be to the advantage of a truly nocturnal animal. It is not surprising, therefore, that tapeta are absent from the eyes of the many occasionally nocturnally-active bird species discussed above. However, it is remarkable that, among the truly nocturnal birds, tapeta have been reported from only a few species of Caprimulgidae (Nicol and Arnott 1974), but it does suggest that in these particular birds vision at night is important.

The second piece of evidence that vision may be important in nightjar behaviour comes from field observation of the foraging tactics which birds sometimes use. Schlegel (1967, cited in Cramp 1985) recorded that nightjars normally approached large flying insect prey in flight from below or, less often, from the side. Only occasionally did the birds swoop down to flying insects. This can be because they are better able to detect insect prey in silhouette against the sky rather than against the darker background of vegetation, something which we can easily verify for ourselves under similar circumstances. If vision was not important and some other cues were sufficient to guide birds to their prey, it might be expected that it would be approached as often from above as from below.

A third piece of evidence that nightjars may be guided by visual cues is more tenuous, but also instructive. It is recognised that where nightjars are flying at night, the birds' response to a human standing still is more or less random and that the bird may show no special interest. However, if a large white handkerchief is waved silently above the head, a nightjar is likely to be attracted and will apparently investigate the waving flag from relatively close quarters, sometimes momentarily hovering. This suggests two things: first, that the bird is at least tempted to investigate a conspicuous visual signal but, secondly, that the bird is relatively easily confused. Presumably the bird mistakes the handkerchief, either for another nightjar [white patches on the tail and wing feathers are used conspicuously in display, Cramp (1985)], or for possible prey such as a large moth. Whatever the source of the confusion, the bird clearly is not able to detect the decoy handkerchief for what it is until at close quarters. Other sources of confusing stimuli which are reported to elicit close approach include torchlight and sticks thrown into the air (Cramp 1985).

Finally, mention should be made of the possibility that Caprimulgidae may use echolocation for the detection of prey. This was specifically suggested for the Common Nighthawk *Chordeiles minor* which is common in open habitats, including suburbs and towns, through most of North America. Griffin (1958) proposed that this bird might use echolocation, but there is no evidence that any bird species, other than those which breed in caves, is capable of echolocation. Also, as explained when discussing those few species (Chapter 6), there are good reasons to believe that in any bird the spatial resolution of the echolocation employed is unlikely to endow it with the ability to detect even large insect prey of the kind sometimes taken by nightjars and nighthawks.

NOCTURNALITY AMONG THE OWLS (STRIGIFORMES)

In many cultures the association of owls with night-time is so instantly understood that a story teller, dramatist or poet can refer to the call of an owl and be sure that the audience will immediately know that the narrative is set in night-time. It might even be true that the use of an owl as a symbol of night-time is one of the most powerful literary symbols drawn from the natural world. Everybody "knows" that owls are nocturnal. However, despite its almost universal acceptance this is only a half-truth, for of the 135 species of owl presently extant in the world, probably less than half, and maybe as few as 30%, are strictly nocturnal within the definition used here.

Burton (1984) presents the most comprehensive review of the owl species throughout the world and analysis of the summarised data presented there suggests that probably only about 40 species actually fall within the strictly nocturnal category. A number of species are more or less diurnal, some may be crepuscular and diurnal, others crepuscular and nocturnal. Some species would seem to be somewhat flexible in that certain individuals or pairs, perhaps in certain habitats, seem to take on a strictly nocturnal mode of life while others are more crepuscular or even diurnal in their activities.

Detailed data on the occurrence of nocturnality in many of the world's species is, however, not available. Many species live in remote areas, often in dense woodland, where systematic studies have not been possible. As was noted above, a bird once thought to be strictly nocturnal, the Kagu *Rhynochetos jubatus,* has been shown to be primarily diurnal, and it is quite possible that ideas on the extent of nocturnality in some owl species will need to be revised as the result of future detailed studies. The paucity of knowledge about owls in many parts of the world was exemplified by the discovery, as recently as 1981, of a hitherto unknown species in the forests of Peru, the Cloudforest Spotted Screech Owl *Otus marshalli.* Equally important when considering nocturnality, is the degree to which anecdotal field observations may have ignored the marked differences in light levels which naturally occur at night (Chapter 2).

Although owls are among the most instantly recognised group of birds, they are very diverse in many aspects of their biology, anatomy and natural history. The Strigiformes are divided into two families, Strigidae and Tytonidae (Voous 1988). The Tytonidae (barn and grass owls) contain just ten species which are relatively uniform in size, structure and habits. The Strigidae (typical owls), however, are a rather diverse group containing birds which vary greatly in size and in many aspects of their biology. Some species such as the Hawk Owls of the genus *Ninox* (Fleay 1968) look superficially more like falcons than the more well-known owls of the genera *Strix, Otus,* or *Bubo.* Owls may be found in a diversity of habitat types and climates, ranging from Arctic tundra to tropical rain forests; some are highly sedentary, others

The Great Horned Owl *Bubo virginianus* has the widest distribution of any owl in the Americas. It is found from Tierra del Fuego to Alaska in a wide range of woodland habitats including boreal forest, tropical rainforest and mangroves.

nomadic, while some are regularly migratory. Some owl species hunt on the wing, others hunt from perches; some are dietary specialists while others are known to be very catholic in what they will eat; prey ranges from moderately sized mammals and birds to fish, insects and soil invertebrates.

It is not the intention here to review in detail the natural history of owls throughout the world. There are already a number of works which do this and bring together references to a large number of detailed studies; for example, Clark *et al* (1978), Bunn *et al* (1982), Mikkola (1983), Burton (1984), Cramp (1985), Kemp and Calburn (1987), Voous (1988). Also, certain owl species, especially those of Europe and North America, have been the subject of detailed long-term studies which have produced a mass of data on the biology and natural history of certain species, and thus their biology must be regarded as well known and their degree of nocturnality reasonably well established. Most notable among such studies are those of the Tawny Owl *Strix aluco,* Southern (1970), Wendland (1984), Hirons (1985); the Short-Eared Owl *Asio flammeus,* Clark (1975); Tengmalm's Owl *Aegolius funereus,* Korpimaki (1981); and the Barn Owl *Tyto alba,* Bunn *et al* (1982).

The aim here is briefly to draw together (mainly using data from the above references) some common themes concerning the biology and natural history of owls which appear to be correlated with differences in the degree of nocturnality. These themes include habitat type, territoriality, dietary spectrum and prey capture technique. They will be drawn together at the end of the chapter to describe a "nocturnal syndrome" in owls. This will provide a background for considering in the light of knowledge on the sensory capacities of owls (Chapter 8), how it is that certain owls are able to complete all aspects of their annual life cycle during the hours of darkness.

Nocturnality and habitat type

One theme which has recurred throughout this book has been that nocturnal activity in birds, especially if it involves flight, takes place in open, spatially simple, habitats. This applies not only to occasional nocturnal behaviours, but even to the regular nocturnal foraging of Nightjars and Stone Curlews. Such a proposition would seem, in principle, quite reasonable since nocturnal mobility is always likely to be difficult, and the fewer the obstructions the more readily a bird, whether on foot or in flight, can move through its habitat in safety.

It is surprising to find, therefore, that among owls this relationship between nocturnal activity and an open type of habitat is not always preserved. Thus, some strictly nocturnal owl species which make use of complex woodland habitats for breeding and roosting, forage in more open areas, either outside the woodland or along rides and in clearings. However, there are a few species which both breed and hunt regularly within the darker and complex conditions beneath a tree canopy. Furthermore, those owls which inhabit the completely open and treeless habitats of tundra, heath, grasslands, marsh and mire, tend *not* to be strictly nocturnal, preferring instead to hunt during the day or twilight only.

Three main relationships between habitat preference and the degree of nocturnal–crepuscular–diurnal activity among owls can be discerned. They are illustrated here mainly by example of species from the Palearctic region.

(1) *Diurnal–crepuscular activity and open habitats*

This relationship among owls is illustrated by three species: the Snowy Owl *Nyctea scandiaca*, the Short-eared Owl *Asio flammeus* and the African Marsh Owl *A. capensis*. These species typically begin their foraging during day-light or twilight. Activity may continue into the night but will often cease shortly after dusk. All three species inhabit throughout the year flat or gently undulating open country practically devoid of trees or other obstructions. The Short-eared Owl prefers open grassland and heathlands throughout much of the northern temperate zone and in South America; while the Snowy Owl inhabits Arctic tundra and open grasslands of the northern hemisphere; the African Marsh Owl, as its names suggests, lives in open marsh and grassland areas throughout much of Africa.

(2) *Diurnal–crepuscular activity and woodland habitats*

This relationship is illustrated by such species as the Hawk Owl *Surnia ulula*, the Great Grey Owl *Strix nebulosa* and the Pygmy Owl *Glaucidium passerinum*. All three prefer extensive woodlands and forests for breeding and roosting but seem to require access to clearings and glades, or to areas of moor and bog, for hunting. The Hawk Owl is regarded as a more-or-less exclusively diurnal species whereas the Great Grey and Pygmy Owls are more likely to hunt at twilight. However, none of these species hunts regularly at night.

(3) *Nocturnal–crepuscular activity and woodland habitats*

Species which exemplify this relationship can be subdivided. (i) Those owls which are more likely to hunt beneath a woodland canopy and are associated with extensive woodland cover. They include such species as the Tawny Owl *Strix aluco*, Ural Owl *S. uralensis*, Barred Owl *S. varia* and the Spotted Owl *S. occidentalis*. (ii) Those species which would seem rarely to hunt under a woodland canopy, but for breeding and roosting, use woodlands, small copses, hedgerows, shelter belts and plantations. They usually hunt over adjacent more open areas. Examples of such birds include Tengmalm's Owl *Aegolius funereus*, Long-eared Owl *Asio otus*, Eagle Owl *Bubo bubo*, Great Horned Owl *B. virginianus*, Little Owl *Athene noctua*, Scops Owl *Otus scops* and the Barn Owl *Tyto alba*.

The actual nature of the open country that the birds may forage over can be quite varied and range from farmland with scattered trees, copses, derelict land and abandoned fields (e.g. Little Owl, Scops Owl and Barn Owl), to larger blocks of woodland with adjacent relatively uniform areas of moors,

The Scops Owl *Otus scops*, Europe's only regularly migratory species of owl. Some populations remain all year on the fringe of southern Europe but most populations winter south of the Sahara.

bogs and clearings (e.g. Long-eared Owl), and to more rocky and dissected terrain, in the case of the Eagle Owl. Thus, in effect, the areas in which these birds spend their lives is a mosaic of open and more closed habitat types, with the majority of nocturnal hunting taking place in the more open areas.

Not all birds of these species are strictly nocturnal wherever they are found. Some, like the Tawny and Eagle Owls, may occasionally be seen abroad during day-light, but the Little Owl and Barn Owl may show a more flexible behaviour which may be in response to particular local habitat characteristics. This is perhaps best exemplified in the case of the Barn Owl.

The Barn Owl has one of the widest distributions of any bird species in the world, with over 30 subspecies now recognised. Bunn *et al* (1982) have described in some detail the variations in nocturnal–crepuscular–diurnal activity of individual birds and pairs. They also describe how some birds within their study area could be described as "woodland" birds, in that they used small woodlands for breeding and occasional hunting, although most hunting took place over adjacent open country or in forest rides, whereas other Barn Owls readily use sites among farm buildings or isolated trees, and it is perhaps these birds which are less strictly nocturnal.

Nocturnality, territoriality and a sedentary lifestyle

The idea of a territory is a familiar one to all ornithologists and bird-watchers. However, this idea is used to describe many different behaviours which share in common one simple factor, the defence of an area by a bird. Thus to be "territorial" a species needs only to defend, at some time in its annual cycle, some physical space from other animals. It will be argued here that territoriality achieves its most extreme possible form in the nocturnal owls. Not all owl species show such a high degree of territoriality, and it seems likely that recognition of these differences in the degree of territoriality among owls is crucial to understanding the basis of their nocturnal activity.

To give some perspective to this it is worth mentioning briefly some of the many forms that territorial behaviour can take in birds:

A Guillemot *Uria aalge* may defend from other Guillemots a territory that is no larger than a small area of cliff ledge, or a flat rock surface, as a site on which a single egg is laid. This area may be as small as 0.05 m², which is probably the smallest defended territory of any bird species (Nettleship and Birkhead 1986).

A Hummingbird may defend an area, sometimes as small as a single plant, which contains nectar-bearing flowers, not only from other Hummingbirds but also bees (Wolf *et al* 1976).

A Blackheaded Gull *Larus ridibundus* may temporarily defend from other gulls a "mobile territory", little more than a few square metres in extent, which contains a small number of foraging Lapwings *Vanellus vanellus* from which the Gull wishes to steal food (Kallander 1977).

Other more familiar examples of territorial behaviour are the breeding and non-breeding territories of passerines such as the Robin *Erithacus rubecula*, Blackbird *Turdus merula* and Great Tit *Parus major*, in which the defended area, where most foraging also occurs, may be in approximately the same location throughout the year (Lack 1965; Perrins 1979; Snow 1988). However, there are also examples of various species of Old World warblers (Sylviidae) which may defend a breeding and foraging territory during the European summer, and then defend a different feeding-only territory in its non-breeding quarters many thousands of kilometres away in Africa.

It can be seen from these example that a bird may defend an area for a number of different reasons and that there is no one function for a territory. Often, the defence may involve food resources, but this is not always so, as for example in the defence of breeding sites in the Guillemot or even the defence of sites used exclusively for mating, as in the case of species whose breeding system involves a lek, such as Ruff *Philomachus pugnax*, Hogan-Warburg (1966), Van Rhijn (1983). Even if food resources inside a territory are defended from other birds, this does not mean that a bird will feed exclusively within that territory and thus be totally dependent upon the defended area for its survival, either for part or whole of the annual cycle. For example, some raptorial birds may defend a territory around the breeding site but nevertheless hunt over a much wider "home range" which may be shared with other birds of the same species (see Brown 1976b and Newton 1979 for example of the different forms of territorial behaviour in diurnal birds of prey).

The above examples of different forms of territorial behaviour, which is by no means exhaustive (see also Howard 1920, Hinde 1956, Wilson 1975, Davies 1978, for further examples and discussions of the idea of territory), does illustrate very clearly the diverse forms that territoriality can take amongst birds. Although the examples refer to a diversity of bird species, even within closely related groups different forms of territoriality have been recorded. This is particularly true among the owls where, although probably all species defend a territory during the breeding season, the species differ markedly in their use of a territory outside of the breeding season. What is most intriguing, however, is that the degree to which owls defend a territory throughout the year appears to be correlated with their degree of nocturnality. It seems that the most highly nocturnal birds are also the most strongly territorial and sedentary species, while the diurnal–crepuscular species may defend a territory during the breeding season only.

Strictly territorial and sedentary owls

The Tawny Owl *Strix aluco* is the most well-studied owl species as regards its territoriality (Southern 1970; Hirons 1976, 1985; Hardy 1977; Wendland 1984). This owl is territorial throughout the year and territorial boundaries alter little, if at all, during a bird's life time, the defended area supplying all of its food requirements. Also, both male and female share the same territory throughout the year, although there is some evidence that females occasionally move territories. Possession of a territory seems essential, not just for successful breeding, but also for the very survival of each individual during the annual cycle. Approximately 60% of young birds die each year, many of starvation, and it is believed that most do not hold territories. Hirons *et al* (1979) have argued that starvation in young Tawnies is directly attributable to this factor. Mortality amongst territory holders is less than 15% per annum and, once established in its territory, an individual owl is likely to remain in that defended area all of its life.

In short, the Tawny Owl is an example of a bird which never goes anywhere in its life except within the bounds of its territory. In the preferred habitat of broadleaved woodland, in Southern England a territory may be as small as about 12 ha (approximately equivalent to a square with 350 metre sides). Further north within Britain, territory size may increase to more than three times this area and it has been argued that territory size is dependent upon the annual average abundance of prey in the area (Hardy 1977; Hirons 1985). Tawny owls are relatively long-lived birds with an average life expectancy in excess of five years, while the oldest known individual in the wild lived 21 years 5 months (British Trust for Ornithology, BTO News, no. 162, 1989).

Territorial boundaries are vigorously defended against neighbours and any intruders are chased through the territory and expelled from it. Such intruders are usually first-year birds who roam the countryside after being expelled from their parents' territory in early autumn. A great deal of the vocal activity of tawny owls heard at this time of the year concerns the ejection of these young birds as they pass through occupied territories.

One further interesting aspect of this territorial behaviour is that although birds vigorously defend their territory against neighbours and intruders, they make little attempt to expand into an adjacent territory even if it becomes vacant. This is true even though the abundance of prey within the territory may change markedly and become insufficient for successful breeding to occur (Southern 1970; Hirons 1985). A previously occupied territory may be left vacant for many weeks or months without any sign of adjacent birds entering into it, with the result that territorial boundaries may survive longer than the birds which they divide (Southern 1970). This stability of boundaries is in marked contrast to the situation in many passerine birds, where territory sizes may change in response to the death or removal of an adjacent territory holder, or a change in the abundance of prey not just between, but also within, breeding seasons (Krebs 1971; Davies 1980).

No other owl species has received such detailed study as the Tawny Owl. However, summarised data for European, Asian and North American species (Mikkola 1983; Cramp 1985; Voous 1988) indicates that almost certainly the Ural Owl, Barred Owl and Spotted Owl have a similar highly territorial and sedentary life style in which the actual annual survival of individuals, as well as breeding success of a pair, is linked to possession of an exclusive year-round feeding territory, which juveniles must acquire in the autumn or probably die during the winter from starvation. These owl species are thus accurately described not only as highly territorial but as sedentary, too. All of them are strongly nocturnal, with extensive woodlands the preferred hunting and breeding habitat.

Partially sedentary owls
There are other owl species in which individual pairs are reported to maintain a territory throughout the year, and thus are essentially sedentary,

but in their case possession of a territory does not appear to be essential for survival. These include species such as the Eagle Owl, Barn Owl, Long-eared Owl, Little Owl and Tengmalm's Owl. It should be noted that all these species are more or less totally nocturnal but are less dependent upon extensive woodlands than the Tawny or Ural Owls; also that they usually hunt outside, rather than beneath, a woodland canopy. In all of these species a pair may defend a territory during the breeding season only and then move to another area for the winter, where another territory may be established or the bird may continue to wander. In the case of many of these species, individuals may not move particularly far between breeding and non-breeding areas, but some individuals may in fact undertake considerable migrations. Also, although one member of a pair, usually the male, may stay within the breeding territory through the winter, the other member may move away.

A well documented example of this more flexible territorial/sedentary behaviour is provided by the Barn Owl. Bunn *et al* (1982) have described how, in their study area in northern England, Barn Owls may remain in the same territory throughout the year or may move to different areas, while at the same time it is known that Barn Owls of Continental European origin may even, occasionally, cross the North Sea to winter in eastern England. The Long-Eared Owl would also seem to show a similar pattern but with more extensive and regular migratory movements, for example, across the North Sea into Britain. The Eagle Owl, although having records of highly sedentary individuals in some parts of its range (one pair having used the same nest site for 30 years, Cramp 1985), is thought to be considerably nomadic in other parts of its range within the USSR. The Little Owl may also show a pattern of individuals or pairs holding a territory throughout the year (in one case for up to four years) while others may move away from the breeding territory during the winter. The Scops Owl which breeds in southern Europe is essentially nocturnal and prefers small woodland with mature timber for breeding and roosting but feeds in more open areas close by. However, Scops Owls are primarily migratory in their habits, leaving Europe for wintering grounds south of the Sahara. The one feature which does distinguish this species, however, from the others discussed here, is that it feeds almost exclusively on large insect prey, particularly crickets and beetles, taken from the ground in open areas, a similar diet in fact to the Stone Curlews.

Non-sedentary owls

The final group of owls to consider are those which regularly defend a territory during the breeding season but may make regular movements away from breeding areas, sometimes becoming nomadic and gregarious, outside the breeding season. Examples of species which fall into this cagegory are those species which may also be classified as more diurnal and crepuscular in behaviour. These include the Snowy Owl, Pygmy Owl, Hawk Owl, Great Grey Owl and the Short-Eared Owl. However, it should be noted that, even in these species, individuals may remain within the same area from year-to-year, depending upon geographical location, but whereas males may remain sedentary, females often move away and the pairs break up (e.g. Pygmy

Owl). Also, all of these species are regarded as potentially eruptive in that the whole population may desert an area and disperse in search of food, with individuals being present in locations where they have not been noted for some time.

The case of the Short-eared Owl (Clark 1975) provides a particularly instructive example of the pattern seen in non-sedentary owls, especially when compared with the highly sedentary nocturnal Tawny Owl. The Short-eared Owl male defends a breeding territory which supplies most, but not necessarily all, of the food for the breeding pair and young. The size of the area defended may change during the breeding season and the average sizes of territories in a given locality are also likely to change from season to season depending upon the abundance of food supply. It seems likely that the breeding pair stay together for the duration of a single breeding season only. Depending upon geographical location, the birds are likely to migrate or to wander in a nomadic fashion to a wintering area where they occupy a similar habitat type to that used for breeding. In the wintering area the birds may remain solitary or start to roost communally, but they hunt alone, sometimes defending an individual hunting territory. However, the birds may continue to wander and even the large communal roosts may be transitory. As the next breeding season approaches, individuals may stay to breed in their wintering area if food is abundant, but most migrate and it seems unlikely that birds return to previous breeding areas or pair with a former mate.

Basically this same pattern is followed even by the Great Grey and Hawk Owls, which inhabit essentially wooded habitats compared to the open grasslands favoured by Short-eared and Snowy Owls. It seems that in all of these non-sedentary species the actual movements are determined primarily by fluctuations in the local abundance of their prey. All of the species breed at relatively high latitudes where the preferred small mammal prey is subject to population fluctuations or regular cycles.

These three patterns of relationships between the degree to which an owl species is nocturnal, its habitat preference and the extent to which it is sedentary can probably be applied to most owl species (at least within Mediterranean, Temperate, Boreal and Arctic climatic zones). However, for the majority of species, data comparable to that available for the Tawny and Short-eared Owls is unfortunately not available.

Nocturnality and dietary spectrum

The diet of all owl species consists of animal prey, and for some the diet is known in great detail from analyses of prey remains in owl pellets. By the same means the diet can be determined throughout the annual cycle. Pellet analyses have even been used to establish likely hunting ranges by relating prey items to the areas where they may be more or less exclusively available.

The actual composition of owl diets varies greatly, depending on the size of the species and its geographical location. Medium to large-sized species tend to specialise in vertebrate prey of all kinds including mammals, birds, fish and amphibia. The smaller owl species may take a similar range of prey items if they are not too large, but they are more likely to depend upon invertebrates, such as earthworms, beetles and grasshoppers. Of particular interest, however, are the dietary specialists, hunting a narrow range of prey types almost exclusively, while others accept a wider range of prey items throughout the year.

While these differences in dietary spectrum no doubt reflect the diversity of prey that is available within a species preferred habitat type, it nevertheless leads to the situation where a relatively large predator takes at some time during the annual cycle, some very small prey items, which must represent relatively inefficient foraging. Again, this difference in dietary spectrum can be correlated with differences in nocturnality and habitat preference. Two species will suffice to exemplify this difference. First, illustrative of one end of the spectrum, is the Short-Eared Owl, concentrating almost exclusively on voles (Microtinae). Secondly, at the other end of the spectrum is the Tawny Owl, whose individual diet may include a range of small mammals up to the size of young rabbits and moles, birds of many sizes, frogs, fish, earthworms and beetles. The difference in dietary spectrum of these two owls would seem to reflect the degree to which they are sedentary. It seems that because the Tawny Owl forages exclusively within its territory, during the annual cycle it must accept a wide range of alternative small prey items when the relative abundance of larger prey declines. However, the Short-Eared Owl, which is of similar body weight to the Tawny Owl, is a specialist feeder whose response to a shortage of prey is not to change its diet but to forage further afield, often making long-distance migrations or engaging in nomadic wanderings. There is a similar response from the Snowy Owl which, in the tundra, subsists almost exclusively on lemmings and voles.

It should be noted that a wide dietary spectrum occurs in most of the more sedentary species and that this even extends to the *Ketupa* and *Scotopelia* fish owl species which might be assumed, from their name, to be dietary specialists. In fact, they take a wide range of prey throughout the annual cycle, although fish does form a regular and substantial part of their diet. It is also known that both birds are highly nocturnal and sedentary.

Nocturnality and prey capture technique

The prey capture techniques of various owl species have been studied in some detail. The strictly nocturnal birds such as the Tawny and Tengmalm's Owls appear to rely almost exclusively on a "perch-and-pounce" hunting technique in which the bird waits at a fixed perch and then drops or swoops on nearby prey, which is invariably caught with the feet. They rarely, if ever, hunt on the wing.

Perch-and-pounce hunting is not exclusive to the strictly nocturnal owls and has been recorded for most other owl species, including those which may hunt in full daylight. Even the Short-Eared Owl and the Snowy Owl which favour open habitat will use this technique if a suitable vantage point is available. The most frequently used hunting technique of the Snowy Owl is a wide visual scan from a low vantage point, such as a tussock or rock outcrop, and when prey is seen to fly close to the ground towards it, even though it may be over one hundred metres away.

Hunting on the wing by slow quartering or coursing is used by many owls which forage over open habitats. Species such as the Short-Eared Owl, Long-eared Owl and Barn Owl may hunt by regular, slow, quartering flight just one or two metres above the terrain. They are sometimes able to hover, or at least hold a stationary position against the wind, and it seems that few if any owls pursue prey on the wing in the manner of a hawk or falcon.

In Europe the Barn Owl *Tyto alba* can frequently be seen quartering for prey over open habitats in daylight.

The quartering flight behaviour of the owls in open habitats is rather like an aerial version of perch-and-pounce hunting, in that prey is taken from an aerial "perch" rather than by active pursuit in flight. This even applies to those owls which have a more hawk-like appearance, such as the Hawk Owl *Surnia ulula*. Hawk owl species of the southern hemisphere, which have long wings and tails more reminiscent of a hawk than an owl, such as those of the genus *Ninox,* take prey both from the ground and from within trees (especially arboreal mammals), but again there seems to be little evidence of pursuit on the wing. Even when owls capture other birds it seems likely that they do so by taking the prey from its roosting perch, or as it feeds, rather than in full flight.

The "nocturnal syndrome" in owls

A number of generalisations have been made above in an attempt to provide a framework for viewing the nocturnal behaviour of all owl species. Although the generalisations are exemplified by data from a relatively small proportion of all owl species, it is believed that they are illustrative of general points against which other species can be compared. Certainly, the behaviour of all individuals within a species, or even of populations from different geographical locations, may not fit into these generalised points, but by taking this approach it is hoped that a more general picture has emerged which describes the "nocturnal syndrome" in owls.

In essence there are two main types of nocturnal syndrome, which differ in only subtle but important ways. First, there are those owls which are strictly nocturnal and which hunt regularly beneath a woodland canopy (e.g. Tawny and Ural Owls). Secondly, there are the less strictly nocturnal owls which, although they use tree cover for roosting and breeding, nevertheless concentrate most of their hunting along woodland edges and clearings, and in open areas away from tree cover (e.g. Barn and Long-eared Owls). The essential difference between the two groups lies in the degree to which they are sedentary. In the former group a totally sedentary life style seems essential, not just for successful breeding but for individual survival from year to year. In the second group, individuals and pairs may be sedentary but this does not seem essential for individual survival, nor are pairs necessarily maintained between seasons. The strictly nocturnal/highly sedentary owls have a wide dietary spectrum and hunt almost exclusively by a perch-and-pounce hunting technique. The less strictly nocturnal species tend to take a narrower range of prey and may hunt on the wing as well as by perch-and-pounce.

Those owls which are only rarely active by night may, likewise, be placed in two groups, but with essentially similar characteristics. First, there are those species which favour completely open habitats (e.g. Snowy and Short-eared Owls). These owls use trees and shrubs to provide shelter for roosting in only the harshest conditions. Secondly, there are those species which favour mainly extensive woodland habitats, but again tend to use open habitats for hunting (e.g. Great Grey and Hawk Owls). What these more diurnal owls have in common is that they are non-sedentary, often nomadic and eruptive, and tend to be dietary specialists favouring prey which is subject to marked fluctuations in abundance.

The above summary highlights two surprising findings. First, it might have been expected that those owls which inhabit open habitats largely devoid of obstacles to flight are more likely to be nocturnal. However, this is not the case. Secondly, but equally surprising, the reverse of this, that the owls which live and forage within the spatially complex and potentially darker woodland habitats are the nocturnal species. This is especially surprising in view of the previously discussed preference in other nocturnal or occasionally nocturnal birds for open air space, devoid of obstacles.

How these owls may cope with the particular, sensory problems of nocturnal activity is the subject of the final chapters. It will be seen that the ecological and behavioural variables identified above are perhaps crucial to this question.

CHAPTER 8

The senses of owls

In the previous chapters discussion of the sensory capacities either of occasional or regularly nocturnal birds has had to rely, principally, on knowledge of the anatomy and physiology of sense organs. Evidence from direct investigations of sensory capacities has been rather incomplete. However, this anatomical and physiological data has permitted some understanding of how, for example, petrels and kiwis may forage at night guided by olfactory cues; or how tactile cues may play a role in the nocturnal feeding of waterfowl, waders, skimmers and nightjars; or the possible role of hearing in the foraging of Stone Curlews. While it cannot be claimed that the nocturnal activities in any of these species has been fully explained, the previous discussions do suggest feasible explanatory frameworks which could be tested experimentally.

The assumption underlying all of the examples has been that the olfactory, tactile, taste or auditory cues available to these various species, enable foraging to continue either in the absense of, or guided only by, minimal visual cues. Indeed it may well be that it is the use of these non-visual cues for foraging which has predisposed these birds to extend their activities into the night. Also, the point has been made a number of times that all of the species (especially if they fly, as well as forage, at night) are active in basically open habitats largely devoid of obstacles. Indeed the point was also made that in those passerine birds which are occasionally active at night, their behavioural repertoire at night was very limited and that in some cases there was evidence that activity at night was a less preferred strategy than activity during the day.

The nocturnal activity of owls presents, therefore, a particularly interesting set of problems. Among this group of birds are species which may be strictly nocturnal, conducting all aspects of their life cycle at night. Furthermore, they may choose to fly in spatially complex habitats which may be made particularly dark by shade from a tree canopy. Thus, in these birds the use of open habitats for foraging is neither a necessary nor a sufficient cause for them to be nocturnally active. However, an apparent paradox is presented by the owls of completely open habitats since, on the whole, they are more likely to forage by day than by night.

In attempting to understand the basis of nocturnality in owls it is not necessary to rely solely upon extrapolations of possible sensory capacities from knowledge of anatomy, physiology or from field observations. Fortunately a number of experimental studies of visual and auditory capacities has been conducted in owl species. These studies allow direct comparisons between sensory capacities and measures of the sensory problems which an owl may face within different natural environments, such as the light levels which occur naturally at night in different habitats.

ECHOLOCATION AND OLFACTION

Because these two cues have been said in previous chapters to provide important information to guide the flight and foraging of certain nocturnally active birds, it is important to make it clear that there is no evidence that either echolocation or olfaction are used systematically to guide nocturnal activity in owls. Certainly, when in flight, owls have never been recorded as producing any sounds which could serve an echolocatory function.

The question of whether owls could use olfaction to aid the detection of prey is a little more open. No systematic experimental studies have been conducted on the olfactory prowess of an owl species. However, one study on the anatomy of the olfactory apparatus of the Short-eared Owl has been conducted (Cobb 1960). This showed that the portion of the brain served by the olfactory nerve (the olfactory bulb) is relatively well developed and comparable in relative size to that of such birds as the pigeon, domestic chicken, Mallard and Herring Gull, and more highly developed than in the Starling. Since the pigeon, chicken and Starling are known to be capable of making subtle olfactory discriminations which are of relevance to their natural behaviour (Stattelman *et al* 1975; Clark and Mason 1987), it would not be appropriate to conclude that olfactory cues are irrelevant to owls.

Olfactory cues could aid in the detection of certain prey at close range or, perhaps, in the identification of prey once captured. This may be possible with small mammals, some of which give such strong odours that their presence in an area can be readily detected by the human nose. Smell could also serve in the process of accepting or rejecting a prey item once caught. (Again, experienced human observers can use the characteristic smells of mammals to aid in the identification of species.) No specific experimental studies have been conducted to investigate this problem, but there have been studies (described below) designed to investigate the minimum light intensities at which owls can find dead prey. All of these studies do suggest that dead prey cannot be located in total darkness. This could be interpreted as evidence that olfactory cues are in themselves not sufficient to guide this behaviour. However, such observations are not entirely conclusive; they may indicate only that a bird is not willing to search for prey in total darkness unless motivated or trained to do so. Nor do they rule out the possibility that olfaction is used in the selection of prey once caught.

SENSITIVITY TO INFRA-RED RADIATION

Evidence that the Tawny Owl could detect infra-red radiation and thus could "see in the dark" in a way not open to other animals, including humans, was first published by Vanderplanck in 1934. Since infra-red radiation is perceived by humans and detected by instrumentation as mainly coming from sources of heat, it seemed possible that owls could detect the presence of a prey animal by the heat of the prey's own body. (This is the basis of modern "heat seeking" devices which are used to "see" through smoke or to detect people buried in a collapsed building.) Although this evidence has now been rejected because of the results of further studies (Matthews and Matthews 1939; Hecht and Pirenne 1940; Hocking and Mitchell 1961), the possibility that owls (and other birds) can indeed detect infra-red radiation still has some currency among some birdwatchers and a few academic ornithologists. It is therefore useful to discuss briefly what is known about infra-red sensitivity in owls and other vertebrates. It will be seen that owls, probably, can indeed detect infra-red radiation, but not with sufficient sensitivity to be capable of using it in the way that was once suggested.

In the eye, light is detected by photosensitive visual pigments which are present in the rod and cone receptors within the retina. These pigments absorb radiation and initiate a chain of events within the nervous system, leading to the sensation of vision. The retinas of different species contain different types of pigment, each maximally sensitive to different parts of the spectrum. However, these pigments do not have a strict cut-off in their sensitivity as regards the wavelength of the radiation which they absorb. Rather, sensitivity declines away from a single region of maximal sensitivity. This means that a pigment can detect radiation over a wide band of wavelengths if the energy of the radiation is sufficiently high. Thus, at the red/infra-red end of the spectrum it is not possible to specify exactly a wavelength of radiation beyond which humans cannot see. In normal life, as when dealing with natural stimuli and everyday sources of artificial light and heat, the upper limit of human visual sensitivity lies in the so-called "far-red", which is the wavelength region between 750–850 nm. Thus infra-red radiation is generally regarded as beginning at wavelengths beyond about 750 nm (Lythgoe 1979). However, human observers can in fact detect intense radiation sources well into the infra-red, up to wavelengths of at least 1000 nm (Griffin *et al* 1947). That is, given appropriate conditions, humans can "see" into the infra-red. However, this sensitivity has no function in everyday life because such powerful sources of infra-red radiation do not occur naturally.

In Vanderplanck's (1934) study of infra-red sensitivity in the Tawny Owl, it was shown that the bird could not detect a bowl of meat or a dead mouse inside a lightless room unless the food was illuminated with infra-red light. However, the intensity of the infra-red source was not controlled, the food was simply flooded with infra-red radiation (whose wavelength ranged from 850–1500 nm). It is possible, therefore, that Vanderplanck did indeed show that owls were sensitive to radiation in the infra-red region. However, such an intense source was used that the result cannot be extrapolated to everyday or natural situations, and cannot be used to suggest that an owl could detect a small mammal by the infra-red radiation which is generated by its own body heat. The studies which refuted Vanderplanck's findings simply used infra-red sources of lower intensity more akin to natural sources of infra-red radiation.

In short, the safe conclusion from these studies would seem to be that owls are no more sensitive, but perhaps no less sensitive, to infra-red radiation than humans. It is interesting to note that there are some species of surface dwelling freshwater fish, for example, perch *Perca fluviatilis*, goldfish *Carassius auratus* and tench *Tinca tinca*, whose visual sensitivity does extend usefully into the near infra-red part of the spectrum. This is achieved in these species by visual pigments which are more sensitive to longer wavelengths of light than those found in most other vertebrate eyes, including those of man and birds (Lythgoe 1979). Studies of visual pigments and direct measures of the sensitivity within the spectrum in the Tawny and Great Horned Owl eyes (Bowmaker and Martin 1978; Jacobs *et al* 1987; Martin 1977; Martin and Gordon 1974a), have shown that the visual spectrum of these owls is similar to that of man and other birds, including such diurnal species as pigeons and chickens (Bowmaker 1977; Bowmaker and Knowles 1977).

That vertebrate animals can detect prey by the infra-red radiation emitted from their warm bodies has been shown in snakes of the family Boidae (pythons and boas) and the sub-family Crotalinae (pit vipers) (Barrett *et al*. 1970; Mattison 1986). However, in both these groups infra-red radiation is not detected with the eye but by a pair of special facial "pit organs" situated on the head somewhere near the eye or along the margin of the upper jaw. Each pit acts something like a pin-hole camera to produce a crude infra-red image of the world, and this image is integrated with normal visual input in the part of the brain concerned with vision (Newman and Hartline 1981). In pit vipers it has been shown that very small differences in the intensity of infra-red radiation can be detected and accurately located, such that the snake could use this organ to detect the presence of prey in total darkness (Bullock and Diecke 1956). There is no evidence that any bird species has a comparable organ.

VISION

ABSOLUTE SENSITIVITY TO LIGHT

Casual field observations have long been interpreted as providing evidence that the visual sensitivity of owls is considerably higher than in humans. Indeed, observers still argue, from their own experiences of owls hunting prey in "the pitch-darkness that descends below the closed canopy of any forest at night", that owls must have higher visual sensitivity than humans (Voous 1988, p. 209). However, as noted in Chapter 2, there are considerable problems with such anecdotal comparisons, especially as regards comparability of visual tasks, even in the same situation, and also with verifying that the human eye is fully dark-adapted.

The assumed high visual sensitivity of owls has been correlated with the apparently large size, tubular shape and frontal placement of their eyes, relative to those of other birds. Figure 8.1 shows just how large an owl's eyes may be, compared with those of a diurnal passerine. The position, shape and relative size of the eyes in this owl are typical of other species, and it can be seen that each eye is so large that the whole of the skull of a Chickadee (Tit) would fit inside it. Even in the smaller Tawny Owl the overall length of the eye exceeds that of the human eye (Martin 1982).

ASSESSMENT OF VISUAL SENSITIVITY

Two different types of studies have been used to gain some understanding of just how sensitive the vision of owls might be. One is to measure directly the "absolute visual threshold"; the second is to assess the minimum levels of illumination under which owls can find prey by sight alone. It should be noted that these two techniques do not measure the same thing and that the first (absolute visual threshold) is of greatest value because it readily allows comparison with other species and with data on the anatomy and physiology of the eye. Measuring the minimum level of illumination at which an owl may detect prey poses many problems of interpretation, especially in comparisons with other species where different levels of motivation and different tasks are involved.

Measuring absolute visual thresholds

The *absolute visual threshold* is a measure of the *minimum amount of light that can be reliably detected by an animal.* The light level is described in terms of its *luminance*. This means that it can be compared readily with similar luminance thresholds for other species; and, since the luminance of a stimulus is independent of viewing distance, it can be compared with measures of luminance experienced naturally in different habitat types (see Chapter 2 for discussion of the differences between luminance and illumination measures).

The use of the word "reliably" in the definition of absolute thresholds is important, because when any sensory system is operating close to the limits of

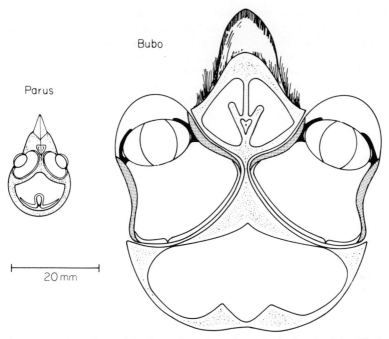

Bubo

Parus

20 mm

Figure 8.1   Drawings of horizontal sections through the head of the Black-capped Chickadee *Parus atricapillus* and the Great Horned Owl *Bubo virginianus*. Redrawn to scale from Wood (1917). In *Parus*, eye shape, size and position are typical of diurnal Passeriform species; in *Bubo*, eye shape and position are typical of other owl species, and eye size is similar to that of the Tawny Owl. In passerines the eyes are fully enclosed within the bones of the skull, but this is not the case in the owls.

its sensitivity the minimum threshold is not clear-cut and must be defined statistically (Pirenne 1962; Baumgardt 1972; Nachmias 1972).

   There are two reasons for defining a threshold statistically. First, in any detector system (whether a man-made device or a sensory system like the eye) there is always random "noise" within the system, which by definition is always fluctuating. In the case of vision, the most important source of "noise" is spontaneous activity within the ganglion cells of the retina. Secondly, the actual emission of light quanta from a source are also considered to be random events both in space and time.

   Fortunately, these random events are of little consequence when the detector system is working well above its threshold (e.g. when the eye is operating at daylight light levels), and the difference between the actual signal from the detector and the noise is very large (i.e. there is said to be a high signal-to-noise ratio) and relatively huge numbers of light quanta are involved. However, as the strength of the signal drops, the background noise within the system does not change and the randomness of the emission of quanta becomes more detectable. Thus the background noise within the system becomes more important and the signal-to-noise ratio is then said to be small. Problems then arise because it becomes increasingly difficult to discriminate between a genuine signal and the fluctuating background noise. As the signal strength decreases still further, an observer eventually finds it impossible to be sure that a genuine signal, rather than noise, has been received. The signal is then said to be below the absolute threshold for detection. It is in the range of signal strengths where an observer is unsure that a signal had been received that the absolute threshold is said to lie. In the case of vision, the observer must decide whether a light (the signal) has been presented or not. Because of the increasing importance of random "noise" close to the absolute threshold this is not an easy task.

Thus, the fluctuating "noise" levels within the visual system give rise to genuine problems for any animal (including humans) whose sensory system is operating close to its threshold. This is because the fluctuations in backgound "noise" make it seem that sensitivity appears to come and go. At one moment we may be able to detect a light, the next moment we cannot, even though conditions have not changed. Thus, in order to specify an absolute threshold, a criterion is adopted which defines the light level which can be correctly detected as a percentage of the number of times the light stimulus is presented. The usual criteria is 75%, because simple guessing will result in a 50% detection rate on most tasks.

To be confronted with light levels close to the visual threshold can be a somewhat unnerving experience. Observers may often be quite unsure that any light is present, but if asked to make a quick decision one way or the other they find that they are correct. At other times an observer may be confident in their decision but find that they are in fact incorrect. This sort of phenomenon is not unique to measuring absolute visual thresholds but applies to all other sensory modalities. It can be easily verified in the case of hearing. If a quietly ticking watch is placed a metre or so away, such that its sound is only just detectable, it will be noticed that the volume of the tick will appear to fluctuate, yet it is known that the sound that it produces is quite regular. This fluctuation in the perceived loudness of the sound is a direct result of the noise fluctuation within the auditory system.

When attempting to measure visual thresholds in an animal, a procedure has to be devised whereby the animal's fluctuating ability to detect light levels close to threshold can be converted into meaningful data that can be compared with the thresholds of other animals including man. To achieve this the science of Animal Psychophysics (Stebbins 1970) has been developed, which depends in the main on the use of experimental procedures in which animals are trained to make specific responses to visual stimuli, which can be presented in a highly controlled manner, so that they can be accurately calibrated. Also the sequence in which various test stimuli are presented can be carefully controlled so that the animal's motivation remains both high and uniform throughout. The animals' motivation to complete the task can also be ensured by extensive training and retraining periods, both before and during the time when thresholds are being tested.

Attention must also be paid to the actual tasks which are compared. It is known in the case of man that many parameters of the test stimulus can in fact subtly alter the final threshold value which is determined (Baumgardt 1972). A small light, briefly flashed on and off, may be detected more readily than a large steadily illuminated light source, but it is the latter type of stimulus which is usually used to define the absolute visual threshold.

Measures of absolute visual threshold in owls

The Long-eared Owl

In the first experimental determination of the absolute visual threshold in an owl, Hecht and Pirenne (1940) concluded that the threshold of the Long-eared Owl was approximately 2.7 times lower than that of man (i.e. the owl was 2.7 times more sensitive). Their procedure was to determine the minimum luminance of a light that would produce a just perceptible contraction of the iris in the bird's eye. It had been previously concluded that a just perceptible contraction of the human iris occurs at a light level approximately 1,000 times above absolute threshold. Applying this same relationship to the owl, and measuring the minimum light level which two human observers could detect, allowed them to calculate the relative difference between owl and human visual sensitivity. There is in fact no evidence that such a relationship between absolute threshold and pupil contraction should, however, apply both in birds and man and, indeed, the actual relationship between pupil contraction and luminance level in any eye is not well understood. Furthermore, Hecht and Pirenne qualified their conclusion by stating that "one cannot put too much reliance on a computation of this kind, which may be off by a factor of 10". Thus very little, if any, reliance can be put on these results, although they do suggest that the difference between human and owl visual sensitivity is not dramatically large.

The Tawny Owl

The only owl species in which absolute visual thresholds have been reliably measured is the Tawny Owl (Martin 1977). With this species a procedure was devised in which birds were trained to sit on a fixed perch in front of two identical translucent panels onto which lights of known intensity could be projected from behind. The birds were then trained to make pecking responses to small metal bars situated in front of each panel. The birds simply had to respond at the panel onto which a light was projected while the other panel remained completely dark. Correct responses were rewarded by the presentation of food. The left–right sequence of the illuminated panel was presented in a random sequence, and the actual light levels used varied in a systematic way. With such a procedure, and with sufficient test trials (the training and testing took many months), enough data could be accumulated to determine at what light level each owl was able to detect the presence of light, correctly, on 75% of the test trials. The same procedure and apparatus were also adapted slightly so that human observers could perform exactly the same task and thus direct comparisons made between man and owls.

Tawny Owl *Strix aluco*

The results of this study showed that the mean absolute threshold in the Tawny Owls was lower than that of two people, tested at the same time, by an average of 2.2-fold (0.34 log units). The human threshold measured with this procedure lay within the normal five-fold range of absolute visual sensitivity found in the healthy human population (Pirenne *et al* 1957), and thus suggested that the humans with which the owls were compared were representative of the human population as a whole. Figure 8.2 shows these results graphically, and it is clear that the range of absolute visual threshold of the Tawny Owl and of man overlap, but that their means differ by about two-fold. Thus, it can be concluded that, even though owls are on average slightly more sensitive than man, we might expect to find individual people with visual sensitivity greater than that of individual Tawny Owls.

Minimum levels of illumination at which owls can detect prey by sight

This approach to understanding something of the visual capacities of owls is of less general value than measures of absolute visual threshold. However, it does have the attraction of being simpler to devise and carry out. The procedure attempts to define a minimum level of *illumination* at which an owl can, apparently, locate dead prey by sight. There are four principle drawbacks with the procedure.

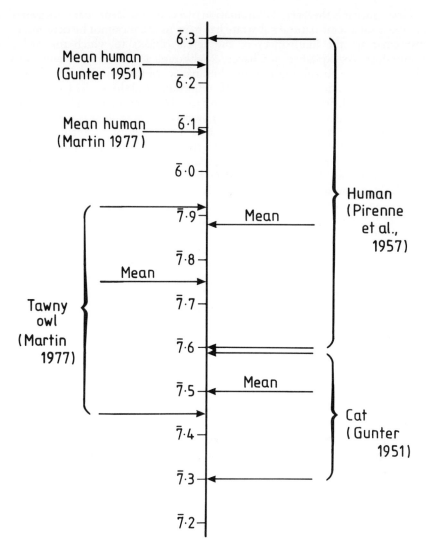

Figure 8.2   The absolute visual thresholds (log cd/m²) of man (Pirenne *et al* 1957), Tawny Owl (Martin 1977) and cat (Gunter 1951). For each species the mean absolute threshold is shown, together with the range of thresholds measured in individual subjects. In all three studies, thresholds were determined in a similar manner using large, uniformly illuminated, patches of white light presented for 15 seconds or more on each trial. In the case of the cat and the Tawny Owl, the absolute thresholds were measured under exactly the same conditions as for humans. These mean human thresholds are also indicated. It can be seen that they fall within the range of human thresholds found in Pirenne *et al*'s more extensive study of human visual thresholds.

First, since it is the overall illumination of a complex scene that is measured (a mouse on a semi-natural substrate has been used), it cannot be determined just what are the luminances of the actual objects to which the owl is responding. As explained in Chapter 2, although a scene may be uniformly illuminated, the actual brightnesses (luminance) of the objects within it may differ markedly, depending upon their reflectance and the angles from which they are viewed. Secondly, the actual task involved is complex and one cannot be confident that a human observer will perceive the stimuli or is motivated in the same way as an owl. Thirdly, it is usually not possible to ensure that the scene is uniformly illuminated. Fourthly, it is not usually practicable to devise a procedure whereby the visual threshold can be statistically defined as described above. Experimenters have been content to define the visual threshold as the minimum illumination at which an owl could apparently recover and eat or move a prey item, rather than a statistical measure of the light level at which it could reliably do so.

Although such studies clearly have shortcomings they have been interpreted as suggesting that the vision of owls may be between 10 and 300 times more sensitive than man. Since these studies are still referred to more or less uncritically in contemporary reviews (Mikkola 1983; Cramp 1985; Voous 1988), it is worth discussing them here.

Dice (1945) determined the minimum levels of illuminance at which four species of owl (Barred Owl *Strix varia*, Long-eared Owl, Barn Owl, Burrowing Owl *Athene cunicularia*) were able to find dead prey. Illuminance levels were not well controlled or measured. The author even admitted that extraneous light was able to leak into the chamber. He stated that, "entering light is very slight and one fails to see any trace of light until he has been in the darkened room for a number of minutes". Given that full dark-adaptation takes 40 minutes (Chapter 2) it is clear that the room was far from completely dark, and Dice stated that "no claim is made that the light intensities at the very weak illuminations used have been determined accurately". Despite this, the results have often been quoted uncritically as though they were definitive.

The experimental procedure was as follows: two dead mice were placed on the sand-covered floor of the chamber, the illumination was adjusted and the owl released into the chamber where it was allowed to search for mice overnight. If the mice had been eaten or moved, a capture was recorded. From estimates of his own ability to detect mice in the experimental room, Dice concluded that the owls were able to detect objects at an illuminance between 100 and 10 times less than that required by him to detect the same object. Since the estimates of illuminance were not accurately measured, the results of this study reveal very little about the vision of owls at low light levels. Dice was actually studying the relative camouflage efficiency of rodents with different coat colours against different coloured substrates, a question which his experimental procedures were better designed to investigate, rather than the relative visual sensitivity of owls and humans.

A further study of this kind was that of Lindblad (1967), in which the minimum illumination level at which owls of five species (Tawny, Ural, Long-eared, Pygmy and Tengmalm's) could approach dead prey, from two metres or more, were reported. The author also compared the birds' performance with his own ability to detect the dead prey in the same circumstances. The results suggested that the Tawny and Ural Owls were over 300 times more sensitive than either the Pygmy or Tengmalm's Owls, or man.

The interesting point about these results is that, if the reported minimum illumination levels at which the Tawny and Long-eared Owls were able to detect prey are translated into the luminance that would be produced on, for example, a piece of white paper (whose reflectance might approximate that of a light-coloured mouse), then these birds' luminance thresholds fall within the range of the Tawny Owl's absolute thresholds reported by Martin (1977) (Figure 8.2). This might suggest, therefore, that both studies had been measuring the same thing. [It is also noteworthy that these levels of illumination are within the range produced by starlight under a tree canopy (see below).]

However, this comparison falls short because if the same considerations are applied to Lindblad's measurements of his own ability to conduct the same task, it is then found that his estimate of human sensitivity is 330 times lower (i.e. 330 times less sensitive) than other more comprehensive measures of absolute human sensitivity to light (e.g. Pirenne 1962). This perhaps indicates clearly the problems of devising a visual task that is comparable for, and will be tackled with equal motivation by, different species.

Although estimates of minimum levels of illumination under which owls can find dead prey may have intrinsic attraction, because of their more "naturalistic" approach, compared with the more artificial setting used for training experiments in the laboratory, differences in the motivation which different animals may bring to the task can clearly influence the results. With training experiments, it can at least be ensured that levels of performance on the task are comparable for all subjects, whether animal or human, before testing begins.

VISUAL SENSITIVITY IN CONTEXT

It may seem surprising, and perhaps disappointing, that absolute visual sensitivity in owls may not be as high as once supposed. However, there are a number of different contexts in which to view these findings, some of which reinforce the idea that absolute sensitivity in man and owls are likely to be similar.

Visual sensitivity in other species

Experiments to determine the absolute visual threshold of an animal using behavioural training experiments are both difficult to design and laborious to conduct. It is not surprising, therefore, that, apart from man, data on absolute visual thresholds are available for very few species. Two particularly pertinent species, with which the Tawny Owl threshold can be compared, are the domestic cat and the pigeon (the only other bird species for which adequate absolute threshold data is available). The cat may be regarded as an example of a mammal which is essentially nocturnal in its habits. The pigeon may be regarded as being strictly diurnal in its habits, and so a comparison between visual performance in a nocturnal and in a diurnal bird is possible.

Absolute visual sensitivity of the owl is approximately 100 times greater than that of the pigeon (Blough 1955; Martin 1977). Therefore, it can be concluded that while the owl may have only a slight advantage in visual sensitivity over ourselves, it clearly has a significant advantage compared to a bird which habitually roosts as soon as light levels approach those of dusk. If the pigeon's absolute sensitivity is representative of other diurnal bird species, then this difference in threshold between them and owls may be sufficient to account for their reluctance to be active in complex habitats, or their inability to conduct efficient visually guided foraging at twilight light levels (Chapter 4). This relatively lower visual sensitivity may also account, at least in part, for the apparently poor visual performance of diurnal birds when migrating at night, which can sometimes lead to their being perceptually confused by artificial light sources.

Figure 8.2 indicates that Tawny Owls are on average about 2.2 times less sensitive than cats (Gunter 1951). However, if account is taken of the range of absolute thresholds found in cats, then it can be seen that the owl and cat populations overlap and that there are likely to be individual owls with higher visual sensitivity than individual cats. One important difference between the cat's eye and those of man and owl is that it contains a tapetum which can serve to enhance sensitivity by a factor of about 2 (Chapter 7).

Visual sensitivity and eye structure

As noted in Figure 8.1, the owl eye is absolutely large and is tubular in shape. Both of these features have been correlated with the nocturnal habit (Walls 1942; Tansley 1965), but just how they were supposed to enhance sensitivity or contribute to visual performance at night was not clear. More recent studies of eye structure in the Tawny Owl have, however, allowed links between its eye structure and visual performance to be better understood (Martin 1982).

One way to view the workings of any vertebrate eye is to see it as being composed of two essentially separate functional systems. First, there is an optical system whose function is to provide a real image of the animal's surroundings. This image is produced on the light sensitive retina at the back of the eye. The retina constitutes the initial part of the second functional system which, together with the brain, analyses the image produced by the optical system. Clearly, the properties of the optical system and of the retina will determine just how an animal perceives its world. A simple but valid analogy is with a video camera, which also contains two systems: a lens system to produce an image, and a system which analyses that image ready for transmission. As in the case of the video camera, the basic optical design of the eye is relatively simple, but its properties can vary greatly and will determine how large the image is and its brightness. Equally the nature of the eye's retina, or image analysing system, will determine just what information it is able to extract from the image, for example the degree of detail that it is possible to resolve.

As in the case of any optical system, such as an eye or a camera lens, its principal properties can be described by just two fundamental parameters, which will be familiar to all photographers. (1) The focal length; this gives an indication of the size of the image produced, the longer the focal length the larger the image of any given object. (2) The f-number; this is a ratio of the focal length of the eye divided by the diameter of the entrance pupil, and gives a measure of the relative brightness of the image that is produced. The smaller the f-number (larger pupil relative to focal length) the brighter the image.

Analysis of the optical structure in the eyes of various species (Martin 1983) has shown that the human and owl eyes differ only little in terms of their fundamental optical properties. Thus the focal lengths of these two eyes are very similar (17.06 mm and 17.24 mm in man and owl respectively) but they do differ slightly in the maximum brightness of the image which they produce. (The f-numbers of human and owl eyes are 2.13 and 1.30 respectively.) The results is that, when viewing the same scene, the image in the owl's eye is approximately 2.7 times brighter than in man. Since the focal lengths of these eyes are the same, this difference arises because the owl has a larger pupil (maximum diameter is 8.0 mm and 13.3 mm in man and owl respectively). The important point about this finding is that the difference in measured absolute visual threshold between man and owl (approximately two-fold) can be accounted for by the fact that the owl eye produces a brighter image when viewing the same scene as a human eye.

This in turn suggests that the retinas in the owl and human eyes are equally sensitive, the essential difference between the two eyes lies in the fact that they produce images of different brightnesses.

There are good theoretical reasons for believing that this should be the case. Although it is often thought that human visual sensitivity at night is poor, and that humans are inadequately equipped to function visually at night, it has been argued that the human retina has attained the absolute limit of visual sensitivity for a vertebrate eye. This limit is dictated by the quantal nature of light and the signal-to-noise limitations of extracting information from an array of photoreceptors which are functioning at the limits of their sensitivity (Pirenne 1962; Barlow 1981). For example, it seems likely that an individual rod photoreceptor in the human eye can respond to the absorption of just a single quantum of light, and that a visual sensation is produced when as few as 80 quanta are received simultaneously at the cornea.

These are absolutely minute quantities of light energy and it is only possible to increase sensitivity by increasing the chances of catching and absorbing more of the quanta that enter the eye. Again, the vertebrate eye seems to be very efficient at doing this. It is estimated that 50% of the energy incident at the cornea actually reaches the retina (Pirenne 1962; Le Grand 1957) and the only way to catch more quanta would be to produce a brighter image in the first place and/or to have photoreceptors which are either larger or contain more photopigment. However, measurements show that the rod photoreceptors of the human and owl eyes are almost identical in all of the parameters which would influence their individual sensitivity (Bowmaker and Martin 1978; Bowmaker and Dartnall 1980) and hence they would absorb a similar proportion of the light quanta that reaches them. Therefore, unless maximum retinal image brightness can be greatly increased over that of the human eye, visual sensitivity in excess of that in humans is unlikely to be found in any owl species.

Thus it can be concluded that human and owl eyes have retinas which are very close to the theoretical limit of sensitivity, and that the only option for enhanced sensitivity lies in increasing the maximum brightness of the retinal image by having a relatively larger pupil (leading to a lower f-number). It seems that the cat's eye has achieved this, compared with the Tawny Owl, but, even in the cat's eye, image brightness is only at the most five times higher than in the human eye, and this difference is sufficient to account for the difference in absolute visual threshold between cat and man (Figure 8.2).

It seems unlikely that any owl species could achieve visual sensitivity greatly in excess of that found in the Tawny Owl. Thus we can consider how large the pupil of the Tawny Owl would have to be if the eye was to have an absolute visual threshold 10 or 100 times that of man. Assuming that the focal length remained the same (approximately 17 mm) these theoretical eyes would require pupil diameters of 25 mm and 81 mm respectively, with lenses correspondingly larger. Clearly, such eyes would have to be exceptionally large as well as peculiar in their general design—quite unlike the typical tubular shape of the owl eye (Figure 8.1), and indeed quite unlike any vertebrate eye so far described (Walls 1942; Tansley 1965; Martin 1983).

That the retinas of the cat and owl, and the peripheral portions of the human retina, can be regarded as specialised for functioning at low (night-time) light levels is suggested by consideration of relative brightness of the image in the pigeon eye. In this diurnal bird the minimum f-number is approximately 2.0, i.e. almost identical to that of the human eye (f-number = 2.13), yet the pigeon's eye is 100 times less sensitive than the human eye. Hence this difference in absolute visual sensitivity must be accounted for by the pigeon having a less sensitive retina.

Visual sensitivity and the night environment

Another context in which to view these measures of absolute visual thresholds in owl, pigeon, cat and man is to consider them in relation to the actual visual problems which the night environment poses. Put simply, is the sensitivity of the owl's eye sufficient to cope with the night environment?

In Chapter 2, the light levels of the night-time environment were discussed in some detail and it was seen that the night is far from uniform. Not only does it vary in length according to latitude and season, but the speed of transition from full daylight to night-time can vary markedly. More importantly for the present discussion, it was shown that light levels during night-time proper are also highly variable, depending upon such factors as latitude and season, and the presence of the moon, stars, cloud cover and vegetation canopy. In short, night-time is not characterised by a uniform light regime but by one which can vary, after the end of civil twilight, over a range of 10,000,000-fold, even in the same locality, depending upon time of year.

It seems safe to assume that visual systems which function at night have evolved to cope not with a constant light level, but with this full range of naturally occurring levels.

Figure 8.3 presents the summarised data on naturally occurring luminance levels of a grass of leaf litter substrate both in open habitats and beneath a closed woodland canopy, with or without the presence of cloud cover (Chapter 2). Also shown are the mean absolute visual thresholds for the pigeon, Tawny Owl, man and cat. A number of important points emerge from this Figure, allowing direct comparison between absolute visual sensitivity and natural light levels. Thus it is possible to determine in which type of habitat, and under which conditions, vision of some kind is possible in each of the species shown.

First, consider the pigeon. It is clear from inspection of Figure 8.3 that even in this diurnal bird some kind of vision at night is possible. Many naturally-occurring night-time light levels are above the pigeon's absolute visual threshold, this even applies to luminance levels produced by starlight alone, as long as the bird is in an open habitat and there is no cloud cover. However, under a woodland canopy, even bright moonlight will scarcely permit any useful vision, and cloud cover will render vision impossible. Thus, at night in woodland, the pigeon's visual sensitivity would seem insufficient to cope at all with most of the visual problems which may be presented, but the bird could gain some guidance from visual cues if it remained in open habitats.

Secondly, consider the cases of man and owl. It is clear from Figure 8.3 that in open habitats, even with thick cloud cover, visual sensitivity is sufficient to give both owls and man some kind of vision at all naturally occurring night-time light levels, and that the luminance of the sky will always be well above the absolute threshold for vision. However, even in these species, entering beneath a thick woodland canopy will, under some nocturnal conditions, expose them to light levels which are below their absolute visual threshold. The specific conditions which render vision impossible occur only during starlight. As long as a moon is present (even if obscured by thick cloud) the visual sensitivities of man and owl are sufficient to render some kind of vision possible even in a dense wood. Thus, an owl hunting outside a wood should always have sufficient light for seeing. It is only beneath a vegetation canopy that visual problems are likely to arise, and even then light levels will frequently be sufficient to permit some visual guidance of behaviour. It can be concluded, therefore, that the vision of the Tawny Owl is sufficiently sensitive to cope (or at least to permit some kind of vision) under most naturally occurring nocturnal conditions, but there will be times when vision of any kind is not possible. Such occasions occur only within woodland, not in open habitats.

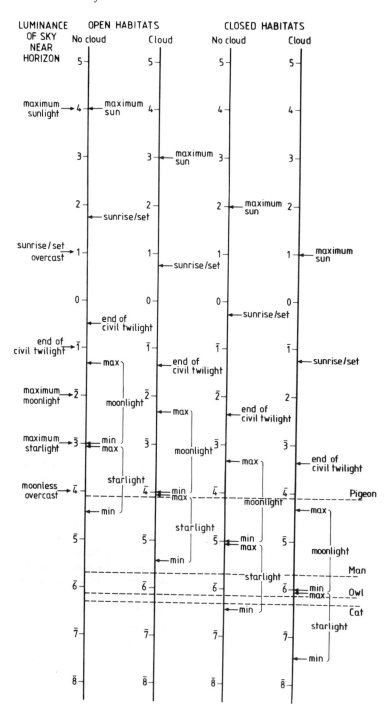

SPATIAL RESOLUTION

Most of the above discussion has considered the absolute minimum of light that an animal can detect. Although this is essential to an understanding of the sensory basis of nocturnal activity (since it defines the lowest limit at which behaviour can be visually guided), it is also important to gain some under-standing of the amount of detail that can be detected at low light levels.

We all know that as light levels fall from those of bright daylight the amount of detail that can be perceived (the spatial resolution of vision) decreases. Small objects or fine detail become impossible to detect, and only large objects can be seen in outline as light levels fall below those of twilight. The reasons for this are now well understood. They lead to the conclusion that poor resolution will occur in *any* eye as light levels decrease (Snyder *et al* 1977; Barlow 1981).

Limits on the degree of detail which can be detected arise principally because of "signal-to-noise" limitations within the retina. Thus, it is not possible to design an eye which would be capable of achieving the same spatial resolution at night-time as during day-time light levels.

While there is data on the visual acuity of the Tawny Owl at high (day-time) light levels (Martin and Gordon 1974b), there is, unfortunately, no infor-mation on the spatial resolution of this bird at low light levels. However, data for the Great Horned Owl *Bubo virginianus* (Fite 1973) is available for a wide range of light levels, as is shown in Figure 8.4. It can be seen that spatial resolution decreases with falling light levels in a similar way to that seen in man. Data on the acuity of the pigeon is more limited but this also is shown in Figure 8.4.

All of this data is from experimental studies using similar tasks which measure "minimum separable acuity". In these studies animals were trained to make discriminations between pairs of black and white (i.e. high contrast) striped patterns presented in different orientations. The width of the striped patterns was systematically altered so that at each luminance level the minimum stripe width detectable could be determined. As with measures of absolute visual thresholds described above, the criterion for defining the threshold was 75% correct performance.

These results show two important points. First, spatial resolution in the owl does indeed decline with decreasing light levels, and that over a wide range of light levels spatial resolution in owl and man is similar. Secondly, at light levels close to the absolute visual threshold, spatial resolution is very poor. For example, Table 8.1 shows the minimum width of an object that would be just detectable at two levels of acuity and at three different viewing distances. It should be noted that these indications overestimate what an animal would be capable of achieving in the natural environment because they refer to the

---

Figure 8.3   The naturally occurring ranges of luminance levels of a grass or leaf-litter substrate which can be experienced in open habitats and beneath a woodland canopy (from Figure 2.5), compared with the mean absolute visual thresholds of the pigeon (Blough 1955), man, Tawny Owl and cat (as shown in Figure 8.2).

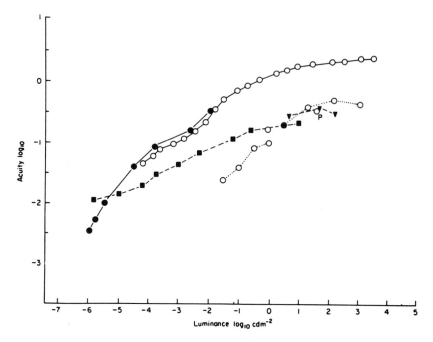

Figure 8.4   The visual acuity of man (●——●, ○——○), pigeon (○ – – – ○, ○P), Tawny Owl (▼ – – – ▼) and Great Horned Owl (■ – – – ■) as a function of luminance. In all of these studies, minimum separable visual acuity was measured. In this, patterns of black and white stripes of equal width are presented and the minimum stripe width which can be reliably detected defines the acuity threshold. The visual angle which this stripe width subtends at the eye is determined (hence viewing distance must be carefully controlled) and acuity is expressed as the logarithm of the reciprocal of this threshold visual angle (in minutes of arc). The higher the acuity threshold value the narrower the just-discriminable stripe width. Data from the following sources: human (Shlaer 1937; Pirenne *et al* 1957); Tawny Owl (Martin and Gordon 1974b); pigeon (Blough 1971; Hodos *et al* 1976; Hodos and Leibowitz 1977).

detection of high-contrast objects (e.g. a black branch against snow) rather than those which more normally occur and are likely to be of low contrast, such as a branch against background foliage, or a small mammal against soil or leaf litter.

It is clear that even quite sizeable high-contrast objects (up to 10 cm across) would not be visible until comparatively close. Certainly, small twigs and branches would not be seen from any appreciable distance. It can be estimated that even an "ideal" high-contrast target, such as a black mouse walking across white snow, would not be visible from further away than approximately 3.5 m even if the owl had an acuity of −2.0 log (see Figure 8.4). To see such a mouse against a darker background the owl would have to be even closer.

TABLE 8.1. The minimum width (cm) of an object (which contrasts highly with the background) which can be just detected at two different acuity levels and three different viewing distances. Acuity is expressed as the logarithm of the reciprocal of the minimum separable angle expressed in minutes of arc (see text and Figure 8.4).

Acuity	Viewing distance (metres)		
	1	2	10
−2.0	2.9	5.8	29.0
−3.5	8.7	17.4	87.0

Thus it can be seen that the amount of spatial detail that can be resolved by any animal, including owls, at nocturnal light levels is considerably lower than during the day-time. Furthermore, it seems possible that spatial resolution in the owl's eye at the lower luminance levels is quite comparable to that of the human eye. However, it is also clear from Figure 8.4 that, as in the case of absolute visual thresholds, the spatial resolution of the owl is superior to that of the pigeon when light levels start to approach those of night-time. Therefore, it can be seen that owls may have a considerable advantage over a strictly diurnal bird in their ability to resolve detail at low light levels, but they do not appear to significantly outperform man under the same light conditions.

EYE SHAPE, EYE SIZE AND VISUAL FIELDS

The tubular shape of owl eyes (Figure 8.1) has often been correlated with the nocturnal life style, as also has the frontal placement of the eyes in the skull. It should be noted, however, that neither the shape nor the frontal placement of the eyes are likely to result directly from the demands of nocturnal vision. It would seem rather that both are a secondary consequence of nocturnality, the evolutionay outcome of squeezing an absolutely large-sized eye (or at least an eye with an absolutely long focal length) into a relatively small skull.

The Tawny Owl's eye is actually longer than that of man. It seems likely that this is the result of attempting to provide maximum image brightness and image size in order to couple high absolute sensitivity with relatively high spatial resolution at low light levels. If a diurnal bird's eye had the same focal length and f-number as the Tawny Owl, but with the usual overall "flat" shape of most diurnal species (e.g. Parus (Tit) species shown in Figure 8.1), the bird's skull would have to be considerably larger than a Tawny Owl's.

Of itself this large eye might not be disadvantageous, but it has been widely argued (e.g. King and King 1980) that one of the prime factors shaping avian anatomy has been the need for weight reduction, especially of the head relative to other parts of the body. Because the eye is basically a water-filled cavity it is comparatively heavy. From this it can be argued that the tubular shape of the owl eye is a product of evolutionary pressure for weight reduction whilst maintaining the overall long focal length of the eye and the relatively large diameter of the pupil. In other words, the tubular eye can be viewed as a more normal "flat" shaped bird eye but with sections trimmed away to reduce overall weight.

The owl eye is, in fact, so large that it cannot be fitted within the protection of the skull, unlike most other birds (Figure 8.1), and there is no room left within the socket for muscles to supply eye movements of any significant amplitude (Steinbach and Money 1973).

However, the eye's tubular shape has consequences for the width of its visual field. It has been shown (Martin 1984a; Martin 1986c) that the optics of the Tawny Owl eye would permit a relatively broad visual field in each eye, comparable to that of the smaller "flat" eyes of diurnal birds such as the European Starling. However, the tubular shape restricts the area of retina which can receive the image and thereby reduces the functional width of the visual field. The effective visual field width in an owl eye is approximately 124°, almost 40° narrower than in the starling and this difference is mainly due to the shapes of their eyes rather than differences in their optical systems (Martin 1986c).

This narrowing of the owl's visual field has consequences for their placement in the skull. Whereas the broader-field starling eye, with its more lateral placement in the skull, can give a relatively wide binocular field, with good coverage of much of the world about it, the narrower-field eyes of the owl need a more forward-facing position to provide binocular coverage of the visual world in front of it, but this is, of course, at the expense of more comprehensive visual coverage of the world around the bird. Thus, in the Tawny Owl, the eyes diverge at an angle of 55° and there is a binocular field of 48°; the total visual coverage extends to 200°. In the Starling, however, the eyes diverge by 114° and total visual coverage can extend to 328°, but a binocular field of similar width to the owl's can still be achieved with the use of eye movements.

The owl could achieve a much larger binocular field, but only if the axes of its eyes were made parallel, rather than diverging, and the greater binocularity would reduce the overall visual coverage of the world about the bird's head. In fact to achieve man's binocular field of approximately 120° (Weale 1960) the Tawny Owl's total visual field would have to shrink to 120° and therefore all vision would be binocular.

Thus a conflict would seem to exist between the demands for an absolutely large-sized eye and the general requirements in birds for a light-weight head. These essentially opposing demands may have led to the evolution of the tubular-shaped eye. Furthermore, the need for some degree of binocular vision as well as wide visual coverage of peripheral areas around the bird has led to the more forward placement of the eyes in the skull. It is certainly clear that the eyes of the owls are not directly forward facing as they are in man and it cannot be assumed that the owls have maximised the extent of their binocular field.

VISION AT HIGH LIGHT LEVELS

It has been assumed by several writers that the nocturnal owls are visually disadvantaged during daylight compared with diurnal bird species. However, there are studies which show that this is not the case. Behavioural training experiments on colour vision (Martin 1974), and physiological studies of spectral sensitivity and retinal structure (Martin and Gordon 1974a; Martin *et al* 1975; Bowmaker and Martin 1978) all show that the Tawny Owl has a fully functional colour vision system, but one perhaps less capable of making colour discriminations as fine as can be achieved by the pigeon (Emmerton and Delius 1980; Wright 1979).

It has also been shown that at high, day-time light levels the visual acuity of owls is very similar to that of the pigeon (Figure 8.4). This implies that during day-time a highly nocturnal owl is not at a visual disadvantage compared with, at least, one strictly diurnal bird species, the pigeon. Both species are, however, inferior in their maximum visual acuity compared with man or the diurnal birds of prey, such as the American Kestrel *Falco sparverius* (Hirsch 1982) and the Australian Brown Falcon *F. berigora* (Reymond 1987). Maximum resolution in these species, as in man, is approximately five times higher than in the Tawny Owl or pigeon at a similar high luminance level.

VISION: AN OVERVIEW

A great deal of information has been presented about various aspects of vision in owls. Most of it refers, unfortunately, to just one or two species but, as explained in the sections on both absolute sensitivity and spatial resolution, fundamental constraints do apply to limit vision in the vertebrate eye and it seems that these birds have evolved vision which is limited by these constraints. This suggests that it is unlikely that any owl species will be found whose visual performance significantly surpasses that reported here for the Tawny Owl.

In the context of discussing how visual performance might be related to the nocturnal habit in owls, two fundamental conclusions are warranted.

First, although absolute visual sensitivity is close to the theoretical limit for a vertebrate eye, and is considerably superior to that of diurnal birds, it is not adequate to ensure vision throughout the range of light levels which occur naturally. However, vision only becomes impossible within closed canopy habitats on starlit nights. In open habitats vision is always possible.

Secondly, although the spatial resolution of owls at low light levels is superior to that of diurnal birds, the actual degree of detail which can be resolved is still not high. Indeed, at light levels close to the absolute visual threshold, spatial resolution is very poor, allowing only the grossest details within the scene to be discerned. Because the sky will always appear relatively brighter than a natural substrate, some detail can always be discerned if viewed in silhouette against the sky.

The overall conclusion from this discussion therefore is that although the vision of owls does show a number of adaptations to naturally-occurring nocturnal light levels these adaptations are insufficient within themselves to account for the nocturnal mobility and foraging of owls under all naturally occurring conditions. This is especially true when species are moving within the darker habitats produced by a woodland canopy.

HEARING

As in the case of vision, it has long been assumed that hearing in owls is special compared with that of other birds. This assumption has rested on anatomical knowledge which showed that the ear openings of many owls species are large and possess many elaborations compared with the simple round hole that forms the ear opening of most birds. Pumphrey (1948) was the first to summarise these differences in ear structure between owls and other bird species, and he described differences in size, shape and the position of flaps of skin and specialised feathers surrounding the actual ear opening. He conjectured on the special properties which these structures might bring to an owl's hearing, but it is only in relatively recent times that detailed knowledge of owls' hearing abilities and their use in their natural behaviour has been published.

HEARING AND PREY CAPTURE

The central importance of hearing in the behaviour of owls was established by the demonstration that the Barn Owl, Barred Owl and Long-eared Owl could capture prey in total darkness, guided only by the sounds produced by small mammal prey as it moved through leaf litter (Payne and Drury 1958). This preliminary work was followed up by a more detailed study of the Barn Owl which also sought to investigate the sensory means by which this ability was achieved (Payne 1962, 1971). A further, less detailed, study by Marti (1974) also indicated that the Great Horned Owl and the Burrowing Owl *Athene cunicularia* could similarly catch live mammal prey solely by sound cues.

Many aspects of this behaviour are worthy of comment, but one often overlooked point about these studies is that, to demonstrate prey capture in total darkness, no elaborate training of the birds was necessary. This suggests that such behaviour is likely to be part of the bird's natural repertoire, and not just a trick that it could be taught in the laboratory. However, it was also found that prey capture by auditory cues alone could not take place in novel situations; the bird had to have experience of catching prey, using visual cues in the same situation, before auditory cues alone would suffice.

Payne's procedure was a simple one which attempted to reproduce a semi-natural setting within the controlled conditions of a laboratory. A Barn Owl was housed in a large, light-proof room with a perch towards each end and a 5 cm deep layer of leaf litter on the floor. Initially, the room was illuminated and the owl fed by releasing live mice into the room. The bird usually captured these mice by dropping down from a perch as the prey moved through the leaf

The stance of a Barn Owl striking at prey in total darkness. The approach to the prey is slower than in the light and the talons are widely spread. This results in enclosure of an area roughly equivalent in shape and dimensions to a mouse's body.

litter. Once the bird had become accustomed to the situation, light levels were reduced and the bird maintained at a low light level (the actual level was not specified) for about five weeks. Eventually the feeding procedure was continued without any light present at all. The bird's behaviour was observed in darkness, using an infra-red imaging device. For three nights the bird did not feed at all but on the fourth night the owl struck the mouse directly and successfully on the first trial. That is, the owl did not seem to need a learning period during which its ability to capture prey in total darkness could gradually improve. Once sufficiently motivated to capture prey in the absence of visual cues, its behaviour was highly accurate.

That the bird was locating and capturing the prey on the basis of the sounds produced as the mouse moved through the leaf litter, was demonstrated by two simple experiments. The first involved dragging a dummy mouse (a wad of paper) through the leaf litter on a thread. This produced only leaf litter rustle cues. In this case the owl captured the paper wad. In the second experiment the floor was cleared of the leaf litter but the mouse had a leaf tied to its tail. Under these conditions the owl "captured" the leaf and the mouse escaped, at least from the initial pounce.

The actual capture of the mouse was studied in some detail. It was found that the owl did not leave its perch while the mouse continued to produce a rustling sound within the leaf litter. In each capture the owl left its perch only after the mouse stopped rustling the leaves. If the mouse actually moved after the bird had left the perch the owl was likely to alight on the ground and wait motionless until a further sound was made, and then it would attempt to approach the sound source again.

It was also shown that the bird approached the prey in a different manner depending upon whether vision was available. In the light, the bird left the perch and approached the prey in a glide, head first, then, moments before impact with the prey, the feet swung up before the face and the talons spread wide. In darkness, the bird flapped all of the way from perch to prey at a speed about half that used during an approach in light. The spread talons were swung back and forth like a pendulum rather than tucked up beneath the tail when making its approach in light. As it neared the prey the owl brought its feet forward until they were level with the tip of its bill. Then, just as in strikes in the lit room, it turned in mid-air, end for end, placing the talons so that they followed in the trajectory formerly taken by the head. Payne (1971) suggested that this was possibly to protect the bird from injury should it hit something in darkness, or if it were to misjudge the distance to the prey.

It was found that the longest distance at which the owl would pounce in darkness was approximately six metres. Payne did attempt to assess the accuracy with which the bird could estimate the distance to the prey, but he concluded that "striking from a perch of the same height for over a year before I tried these experiments, the owl had probably developed an accurate appreciation of where the floor was at any particular angle from the perch" (p. 546).

Direct evidence that owls in the wild might hunt prey, using auditory cues alone, has come from observations of birds taking small mammals from their runs under a continuous snow cover, for example the Great Grey Owl (Tyron 1943; Godfrey 1967; Nero 1980; Hilden and Helo 1981); or likewise, apparently, from runs below a continuous cover of dead grasses, for example the Short-eared Owl (Clark 1975). There has even been a report of a Tawny Owl locating earthworms apparently by the sounds which they produced when disturbing vegetation (Macdonald 1976). Because of the problems of defining complete darkness in the wild it has not been possible to provide definitive evidence that owls take small mammal prey by auditory cues at night. However, there are many suggestions that this might be the case, or that the auditory detection of prey may be commonplace in many species, and many comments to this effect may be found in, for example, Mikkola (1983), Cramp (1985) and Voous (1988).

SOUND LOCALISATION

Although the above section provides ample evidence that hearing may play an important role in the nocturnal behaviour of owls, both the accuracy and the mechanism by which owls are able to locate sounds has been the subject of much investigation [see Knudsen (1980); Kuhne and Lewis (1985) for reviews]. To locate a sound accurately enough for it to be used as the sole cue upon which an owl will base a predatory pounce, requires accurate estimation both of the sound's angular direction from the bird (usually separated in terms of its horizontal and vertical direction from the bird) and its distance. Although the mechanisms and accuracy of angular localisation are well understood, nothing seems to be known about how an owl might judge the distance to an object, accurately, using sound cues.

Ear asymmetry and sound localisation

The key to the mechanism by which owls judge the direction of a sound is thought to lie in the complex asymmetry of the ear openings and of various flaps of skin in front of and behind these openings. The actual openings are positioned just behind the eye and are revealed by parting the feathers at the edge of the facial disc.

The bilateral asymmetry of the outer ears varies markedly across species. A detailed review of its occurrence has led to the conclusion that this asymmetry has no less than five separate evolutionary origins within the owls (Norberg 1977). Ear asymmetry is found in nine different owl genera, but it is not a property of all owl species. It can be a property of the "soft structures" surrounding the actual opening of the auditory canal in the skull, or it may extend to the skull itself. The following describe different examples of asymmetry and is based mainly on the work of Payne (1971), Norberg (1968, 1977, 1978) and Knudsen and Konishi (1979).

In the species of the genus *Asio*, such as the Long-eared, Short-eared and Marsh Owls, the asymmetry of the ears consists of a difference in the vertical positions of the openings on each side of the head, but with the size and shape of openings equal (Figure 8.5).

In the Barn Owl (Figure 8.6) the openings are relatively small and of equal size but the opening of the left ear is higher than the right. Asymmetry is enhanced in this species by big, almost square flaps of skin in front of the ear openings. Figure 8.6 shows that the left flap is placed higher than the right. Also, each ear opening is surrounded by a "facial ruff" of specialised feathers which forms a concave trough (see below for a more detailed description). The ruff on the left is directed slightly downwards whereas that on the right points slightly upwards.

Tengmalm's Owl provides a well-studied example of where asymmetry arises in the skull, rather than in the soft parts of the external ear. In this species the slit-like ear openings in the skin extend the full height of the skull and are symmetrical (Figure 8.7). However, the skull shows marked bilateral asymmetry which is most pronounced in the structure known as the squamoso-occipital wings (Figure 8.8). The heights of the ear apertures in the skull are only about half those of the skin slits. The bony opening on the right is placed about 6 mm higher than on the left. Skull asymmetry has been recorded by Norberg (1977) in two species of the genus *Strix*, the Great Grey and the Ural Owls, but does not occur in the Tawny Owl which exhibits asymmetry only of the soft parts of the outer ears.

In all of these species the bilateral asymmetry of the ears does not extend to the middle and inner ears which are protected within the skull. It is a property only of the outer ears.

Theoretical and experimental studies have shown that this complex asymmetry of the ears serves in the location of sounds. The studies, principally those of Norberg (1968, 1978), Payne (1971), Iljitschew (1974), and Knudsen and Konishi (1979), have shown that the asymmetry modifies sounds as received at the ears. Norberg (1968) measured this by replacing the tympanic membranes (ear drums) with small microphones in a reconstructed head of a

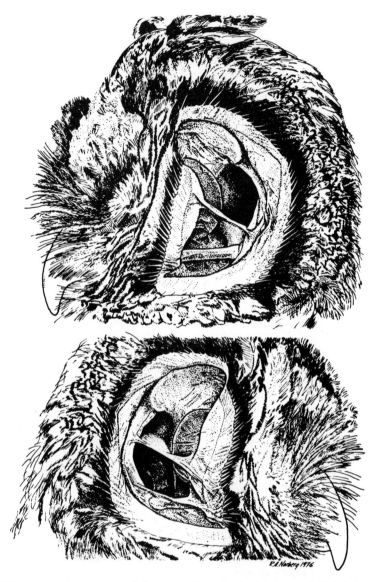

Figure 8.5   Asymmetry of the external ears in the Long-eared Owl *Asio otus*. The head is viewed from the side and the feathers parted at the edge of the facial disc, to reveal the ear openings. The flaps of skin in front and behind the ear openings are displaced, to show the difference in the positions of the ear openings. While the total height and width of the ear openings are the same on both sides of the head, the entrances to the inner ears are in different positions, with the openings on the left side (top), higher than that on the right (bottom). (From Norberg 1977.)

Tyto alba (Scopoli)

R. Å. Norberg 1976

Figure 8.6   Asymmetry of the external ears of the Barn Owl *Tyto alba*. The facial disk feathers have been removed to reveal the asymmetrically positioned flaps of skin in front of the ear openings (the preaural skin flaps). The ear openings themselves are small and almost square, and are concealed behind the flaps. Removal of the facial disc feathers also reveals the feathers of the facial ruff. These lie behind the preaural flaps and almost completely surround the face. It can be seen that they form a concave hollow surrounding each ear opening. (From Norberg 1977.)

dead Tengmalm's Owl. He was then able to record how sounds of different frequencies, presented at different positions about the bird, were altered by the asymmetry in their intensity and time of arrival at each ear drum. Payne (1971) and Iljitschew (1974) conducted similar studies in the Barn and Long-eared Owls, respectively. All studies showed that sounds were altered in complex ways, depending upon the position of their source relative to the head. It was also found that the sounds received at each ear drum became progressively different as the frequency of sounds increased. In the case of

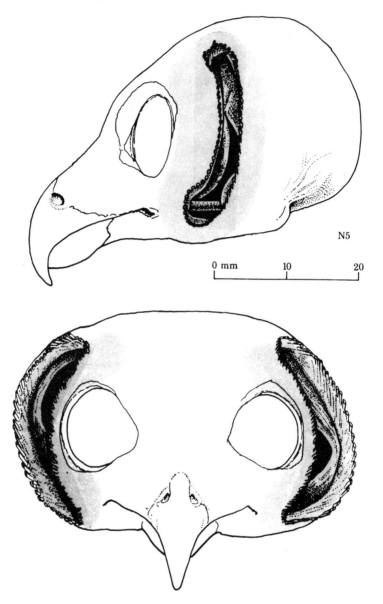

Figure 8.7   Asymmetry of the external ears of Tengmalm's Owl *Aegolius funereus*. The head is intact but the feathers have been removed. The edges of the flaps surrounding the ear openings are indicated by the stumps of cut feathers, but the flaps are shown in their natural position and have not been folded back as in Figure 8.5. The full extent of the outer ear cavity beneath the skin is indicated by shading. The asymmetry in the positions of the openings and also the underlying asymmetry in the skull (Figure 8.8) are clearly evident. (From Norberg 1978.)

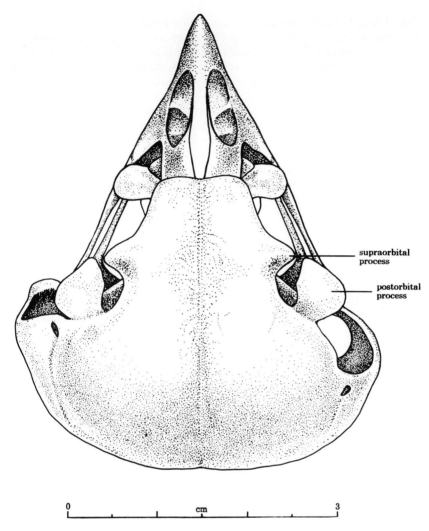

supraorbital
process

postorbital
process

0   cm   3

Figure 8.8 The skull of Tengmalm's Owl *Aegolius funereus* showing the marked asymmetry associated with the external ears (cf. Figure 8.7). (From Norberg 1978.)

Tengmalm's Owl, marked differences were found above about 5–6 kHz (1 kHz = 1,000 cycles per second), and in the Barn Owl differences became most marked above 8.5 kHz.

Various theories have been advanced as to how the way that sounds are received at the two eardrums might serve in the accurate localisation of sounds, especially their role in the localisation of sounds within the vertical plane. In reviewing these theories, Kuhne and Lewis (1985) have suggested that although localisation depends upon modifications of the sound signal by the asymmetric outer ears, different localisation strategies may be employed in different species. They also point out that little is known about the role of controlled movements of outer ear structures, particularly the feathers of the facial ruff and the flaps of skin which surround the ear openings in some species. It has been suggested (Schwartzkopff 1962) that owls may actively control these structures to aid the localisation of sounds, as mammals do with their pinnae when locating sounds. Kuhne and Lewis (1985, p. 269) conclude that "The relative importance of the directional cues arising from external ear asymmetry and ear position control need to be investigated behaviourally, biophysically and physiologically for a number of different species before a full understanding of the mechanisms of sound localisation in the owl can be reached".

The accuracy of sound localisation

Measuring the accuracy with which a bird can locate a sound both in azimuth and elevation is extremely difficult because of the need to establish a suitable behavioural response. This has been overcome in only one species, the Barn Owl. Payne (1971) estimated, from the distance at which an owl could catch a mouse in total darkness, that the accuracy of sound localisation was to within one degree. Knudsen and Konishi (1979) were able to exploit the fact that, like many other owl species, a Barn Owl when sitting calmly on a perch will rapidly move its head to look towards the source of a brief sound. By devising a means whereby the position of the bird's head could be accurately calibrated, Knudsen and Konishi were able to determine the accuracy with which the bird could look towards a source of sound presented briefly at various random locations. The results showed that the Barn Owl could locate the sound of "white noise" (a hissing noise containing sounds across a wide frequency band; similar to leaf litter rustle) to within less than 2° in the horizontal plane and to about 4° in the vertical plane (Figure 8.9). Pure tones were located less accurately, but if such sounds were used those between 7–8 kHz were located most accurately (Konishi 1973b).

However, such accuracy was only achieved if the sound was presented directly in front of the bird. Figure 8.9 shows that performance deteriorated with increases in elevation and eccentricity. Comparable data for other bird species is not available because of the difficulty of finding a suitable means by which birds can indicate where they estimate the source of a sound to lie. However, the limited data that are available [Gatehouse and Shelton 1978 (Bobwhite Quail); Jenkins and Masterton 1979 (Pigeon); Klump *et al* 1986 (Great Tit)] have led to the conclusion that no other birds are likely to match the sound localisation performance of owls (Knudsen 1980). Indeed, the best performer of those species is the Great Tit, which can locate alarm calls to an accuracy of, approximately, only 18° in the horizontal plane.

To give some context to the accuracy of auditory localisation in owls, Figure 8.9 also shows the accuracy with which humans are reported to locate similar sound sources. Clearly, humans can out-perform owls at all locations,

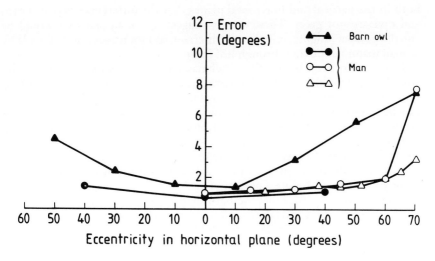

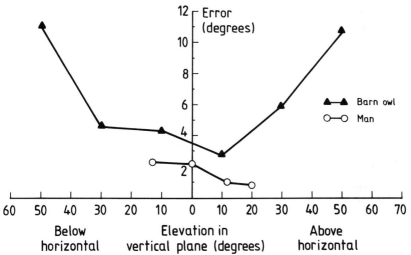

Figure 8.9 The accuracy of sound location in the Barn Owl *Tyto alba* and in man. Location accuracy is indicated by the error in degrees with which the owls could look towards a sound source. In the case of the humans, accuracy was measured by asking subjects to indicate the position of the sound source by reference to a grid system, behind which the sources were concealed. All studies used short bursts of white noise as the sound to be located. In all studies the subjects started with the head facing straight ahead (at 0°) and the test sound presented at various eccentricities, either in a horizontal (top) or vertical (bottom) plane. All data for the Barn Owl from Knudsen and Konishi 1979; data for humans (horizontal) from Schmidt *et al* 1953; Sandel *et al* 1955; Mills 1958; human (vertical) from Roffler and Butler 1968.

both in the vertical and horizontal planes, but the differences between man and owl are not great. These results suggest that man and owl would be equally good at judging, initially, the direction of a sound such as that made by a small mammal moving through leaf litter.

The actual process by which an owl may locate and capture a prey item by sound alone would not depend, however, only upon the accuracy of localisation reported here. A sound produced at the periphery, whose location could not be determined with great accuracy, could be located with far greater accuracy simply by changing the position of the head, and this is indeed what Barn Owls are reported to do when a sound is first presented to them (Payne 1971). By a series of adjusting movements of the head, further sounds from the prey would allow the prey to be brought within the frontal plane where accuracy is highest, before a pounce is made.

There is even evidence that the Barn Owl can determine the direction of a moving sound source and take this into account when pouncing on prey. Thus Payne (1971) showed that the bird would orientate the long axis of its spread talons to parallel the direction in which a wad of paper was dragged through leaf litter. Presumably this would serve to increase the chances that a small rodent would be caught effectively in the talons.

It is perhaps worth noting that the famed, highly flexible, necks of owls may have more to do with sound localisation than the frontal position of the immovable eyes. Because sounds are most accurately located directly in front of the owl's head, this will also mean that by positioning the head to "look" with greatest accuracy at the sound source, the potential prey will also be aligned with the bird's binocular visual field which is centred just above the line of the bill (Martin 1984a). Thus both eyes and ears "look" with greatest spatial accuracy in the same directon, and the flexible neck means that a potential prey item can be brought rapidly into the area before the face where this can be achieved.

Judging the distance to a sound source

For sound to be used as a reliable cue in prey capture it would seem necessary to judge its distance as well as direction. Payne (1971) could find no evidence that Barn Owls could judge/estimate the distance to a sound source accurately if it was not at floor level. Furthermore, the observation that in total darkness Barn Owls approach their prey with the feet always in readiness for an impact does suggest that prey distance is not determined with great accuracy. Humans seem to have difficulty at first in determining the distance to a sound source, especially if it is one with which they are not familiar or is in unfamiliar surroundings (Coleman 1962). However, experience with that sound and situation soon leads to rapid learning and the distance can then be determined with some accuracy. It seems that what is learnt is the way that sound is altered or degraded, in loudness and frequency, as it travels over various distances within the specific surroundings. That is, once the characteristics of the source sound are known and we are able to experience that sound in different situations, it is then possible to estimate its distance from us. Thus, estimating the distance to a sound would seem to rely particularly upon cognitive or learning processes, rather than physiological processes of the kind outlined above regarding the localisation of the angular position of a sound source.

SENSITIVITY TO SOUND

Data on the absolute sensitivity to sounds of different frequencies is available for eleven species of owls (van Dijk 1973; Konishi 1973a). As in measuring absolute sensitivity to light, the absolute threshold for sound has also to be defined in statistical terms, and it is recognised that significant individual differences occur (Stebbins 1970). Since the data for nine of the owl species are derived from single birds it would be misleading to search for interspecific differences in absolute sensitivity. In a review of the available data on hearing threshold in birds Martin (1984b) concluded that by pooling data for all owl species it could be seen that their hearing is, on average, significantly more sensitive than that of any other bird species so far examined [a total of nine species: Budgerigar *Melopsittacus undulatus* (Psittaciformes), Turkey *Meleagris gallopavo* (Galliformes), Pigeon (Columbiformes), and six species of passerines]. However, while the average difference between owls and pigeons is approximately 25 decibels (dB) (equivalent to approximately 300-fold difference in sensitivity) the variability in the measures for these two groups would suggest that there are some individual pigeons which may have their lowest thresholds within the range of the thresholds of individual owls. Thus, Harrison and Furumoto (1971) reported individual pigeons in which lowest thresholds reached −6 dB, while van Dijk (1973) reported three owl species (Scops Owl *Otus scops*, White-faced Scops Owl *Otus (Ptilopsis) leucotis*, and the Forest Eagle Owl *Bubo nipalensis*) with lowest thresholds of −6 dB, −6 dB and −5 dB, respectively.

When the lowest auditory threshold for the owls are compared with those of man and cat, however, it is found that there is no significant difference between them. Masterton *et al* (1969) concluded from their review of hearing in mammals that at the higher phyletic levels within this group (which included 19 primates and three carnivora) the lowest auditory threshold varied little across species. Thus it can be concluded that the hearing of owls, man and the higher mammals is of similar sensitivity. Furthermore, this level of sensitivity is the highest of any vertebrate species yet tested.

It would seem that, as in the case of vision, auditory sensitivity in owls is significantly superior to that of diurnal bird species, but is very similar to that of man and other mammals and has probably reached the absolute limit of auditory sensitivity in vertebrates. Definitive auditory threshold data for humans (Sivian and White 1933) when compared with the available data for owls indicates that there are likely to be individual people whose hearing is superior to that of individual owls, and vice versa.

The facial ruff and hearing sensitivity

The physiological mechanisms or anatomical structures responsible for providing owls with higher auditory sensitivity than diurnal birds are not well understood. However, it seems possible that the facial ruff may be partly responsible. The facial ruff (Figure 8.6) is formed of modified, dense feathers which have a relatively thick rachis and reduced vanes. The ruff feathers are inserted directly behind the ear openings in a special flap of skin. They form a concave wall which almost encircles the face in some species, such as Barn Owls. It is less complete in the Striginae and Buboninae. In the Barn Owl (Payne 1971) the ruff feathers are very tightly packed and form the walls of two approximately parabolic shaped troughs at the approximate centre of which lie the ear openings (Figure 8.6). The facial ruff is covered by the feathers which form the owl's characteristic facial disc. These facial disc feathers do not attenuate or reflect sound because their vanes are very open; however, the features of the ruff do reflect sounds strongly.

From consideration of the dimensions of the ruff it has been calculated that it will act in a manner similar to a parabolic antenna, collecting sounds and roughly focusing them towards the small area of the ear opening. It can therefore act as a sound amplifier. Konishi (1973b) calculated the sound amplification function for the Barn Owl's facial ruff to be about 10 dB for sounds of about 7 kHz frequency, which falls in the frequency range where the Barn Owl is most sensitive. It is also the frequency range used for the most accurate location of sounds. This 10 dB (or 10-fold) enhancement of sound pressure level reached at the ear may, in fact, be sufficient to account for the average difference in auditory sensitivity between owls and diurnal birds.

The facial ruff may thus be an important element in the enhanced auditory sensitivity of owls compared with bird species which lack them. However, since the ruff also forms an important part of the sound localisation apparatus, it clearly has a function beyond the enhancement of sensitivity. A facial ruff is not unique to owls. It has also been reported in Pallid and Marsh Harriers, *Circus macrourus* and *C. aeruginosus* (Accipitriformes), (Iljitschew 1974), and although nothing is known about hearing in harriers it is thought that it may, on occasion, play an important part in their foraging for prey in grassland habitats (Watson 1977).

Limits to hearing sensitivity

Just what might be dictating the lower limit, or absolute threshold, to hearing in owls and mammals is not clearly understood. Certainly the hearing of man, other mammals and owls is very sensitive. Just how sensitive can be understood by considering the various sounds which have been discussed as possible sources of constraint on the absolute sensitivity of hearing in man. These include thermo-acoustic noise produced by the Brownian motion of air molecules upon the tympanic membrane (Sivian and White 1933), self-noise produced by the pumping action of the blood inside the body (Wever and

Lawrence 1954; Diercks and Jeffress 1962) and thermal agitation within the cochlea of the inner ear (Harris 1970).

A more recent suggestion has been that of Martin (1984b) who proposed that naturally occurring ambient sounds could provide a limit to hearing. Ambient sounds occur in all environments, and among the factors which influence the frequency spectrum and pressure levels of this background sound are wind (and the turbulence associated with it as it passes over terrains of differing topography), humidity, the nature of local vegetation, and the presence of animal vocalisations. It was shown that the quietest naturally occurring conditions are experienced on windless nights, which in temperate climes probably occur on ten per cent of all nights. At night, air turbulence, the source of much wind noise, is usually at a minimum due to a more uniform temperature distribution, which is often the result of a temperature inversion. Variability in minimum ambient noise is greatest at low frequencies, where the sounds produced by distant weather systems, ocean waves and air turbulences have their greatest energy (Gossard and Hooke 1975). Winds blowing up to 100 km away can affect the locally occurring sound level at a windless site. Thus there is no such thing in nature as a perfectly quiet night, there will always be some residual background noise, which is likely to have been present throughout the evolutionary history of the hearing systems of terrestrial vertebrates. It seems possible that this minimum background noise could dictate the ultimate limit to hearing sensitivity.

Certainly, measures of this minimum background noise (Martin 1984b) would suggest that if human hearing was more sensitive than it is, we would not be able to detect any more information in the environment, because individual quiet sounds would be masked and rendered inaudible by the naturally occurring minimum background noise. As in the case of natural light levels, discussed in Chapter 2, actually experiencing the minimum sound levels of natural habitats is difficult due to the presence of man-made sounds. Even in remote areas the noise of vehicles, aircraft and ship's engines may still be heard. However, unlike the visual system there is no need to undergo an extensive period of adaptation in order to achieve maximum sensitivity, although exposure to particularly loud sounds, such as a car engine, can raise hearing thresholds for some time after the sounds have ceased. Another difficulty is that the sensitivity of human hearing decreases with age, there being a progressive loss of hearing at the higher frequencies from an early age. Thus, two individuals of different ages are likely to experience quite different sound regimes even when listening to the same natural sound conditions.

HEARING: AN OVERVIEW

In a comparative context the hearing capacities of owls show a number of interesting parallels with their visual capacities. First, there is good evidence that in their absolute sensitivity owls outperform diurnal birds and are equal to mammals, including man. Secondly, this absolute sensitivity is probably close to the ultimate limit which can have evolved, given the physiological and environmental constraints on hearing in a terrestrial environment, i.e. hearing is as sensitive as it can be. Thirdly, owls are superior in their ability to localise sounds compared with diurnal birds but that they do not outperform man. There is no doubt that this localisation is of sufficient accuracy to allow owls to capture noisy prey by sound cues alone, and there is good evidence that such prey capture may form part of the birds' natural repertoire of behaviour. However, the extent to which this may depend upon experience of perch positions and possible flight trajectories within a particular locality is not clear, but it seems likely that prey capture by auditory cues alone cannot occur in novel situations.

CHAPTER 9

The owls' solution to nocturnality

The previous two chapters have presented a great deal of diverse information about the natural history, hearing and vision of owls. In truth, much of that information concerning vision and hearing comes from detailed studies of just a few species, and it cannot be assumed that all of these results apply to all owls. However, the data on vision and hearing do have important general utility. It was shown that in both of these senses the absolute sensitivity found in those owl species investigated to date, is very close to the theoretical maximum predicted by the physical nature of the light and sound, and the physiological constraints on the design of seeing and hearing systems in terrestrial vertebrates. Thus these measures of visual and auditory perform-ance describe a limit which is unlikely to be surpassed to any significant degree in other owl species.

171

In a similar way, the description of the light regimes (Chapter 2) which may be experienced at night in different habitat types, under natural conditions of illumination, are also likely to be widely applicable. The only exception is the possibility that, under some forest canopies, attenuation of the incoming light may be even greater, and hence the luminance levels beneath them even lower than those described in Figure 8.3.

In both vision and hearing it has been seen that sensitivity in owls is extremely high. The visual system can respond to the reception of just a few quanta of light and the ears can detect quite minute sounds which, under most circumstances, are rendered inaudible by the general natural background noises of the natural world. It has also been seen that owl vision and hearing exceed in their sensitivity that of the pigeon, a typical strongly diurnal bird. That is, compared with the pigeon, owls would certainly seem to have significant sensory advantages in the more exacting night-time conditions.

There is, however, one problem in these comparisons. Both in absolute visual sensitivity and spatial resolution, and in auditory sensitivity and sound localisation, owls and man are approximately equal. Such a finding may, at first consideration, be seen as a stumbling block to understanding the basis of nocturnality in owls. People have tended to come to the question of under-standing nocturnality in owls with the preconception that in hearing and vision man is less well equipped and that to be mobile at night any nocturnal animal must *ipso facto* surpass man in its sensory performance, i.e. there must be a "super sense" solution to the problem of nocturnal activity in owls.

This idea arises, perhaps, for two main reasons. First, we are very rarely aware of what we are able to do at night in natural circumstances, armed only with our senses. People tend to believe that they are inadequately equipped to cope with activity at night. This conclusion is often reached without ever testing our own sensory abilities, or becoming familiar with the night and all its various light regimes. Secondly, there is a tendency to assume that a simple "super sense" explanation will be sufficient to account for nocturnal mobility and foraging. Thus we ignore the possibility that the differences in our activities and those of other animals, at night, could lie in the actual tasks undertaken, and in possible behavioural adaptations which permit those tasks to be completed, rather than in simple sensory superiority. Recognition of these ideas may free us from some unhelpful assumptions about our own inadequacy and the owls' superiority when abroad at night.

PREY CAPTURE BY HEARING AND "FLYING BLIND"

The demonstration that owls not only detect, but also capture, prey in total darkness solely by auditory cues is clearly of considerable importance in understanding a number of aspects of their nocturnal behaviour. Clearly it must be concluded that much prey capture, or at least prey detection, in the wild is achieved by auditory cues. Indeed the daylight observations of owls diving through snow to capture prey attests to the importance of auditory cues, even at high light levels. Prey capture in total darkness, using auditory cues alone, appears to be part of the species' natural repertoire of behaviour, rather than a product of elaborate training by an investigator. This suggests that such hunting may occur regularly in the wild.

The above also demonstrates that an owl is willing to fly, at least short distances, without the benefit of visual cues. This suggests that vision is not a prerequisite for foraging and that owls are willing to act "blindly". This finding corresponds with the conclusion that even though owl vision is highly sensitive and close to the ultimate limit predicted for the vertebrate eye, this sensitivity is insufficient to provide vision under all natural conditions. Specifically, Figure 8.3 indicated that, under a woodland canopy, light levels may frequently fall below an owl's absolute visual threshold, and that visual guidance to objects, at least on the floor below a perch, would not be possible.

Consideration of the spatial resolution of the owls' visual system at low light levels also demonstrated that, at light levels close to absolute threshold, only large objects could be detected and that, therefore, visual guidance could employ only gross cues in the environment, not fine detail of the kind that is assumed to guide a diurnally active bird as it moves through vegetation. Corroborative field evidence, that owls cannot always see fine spatial detail, comes from observations that Tawny Owls may fly into branches and even tree trunks if they are surprised by an observer (Hirons, personal communication). Furthermore, in a survey of the occurrence of natural bone breakages in birds, owls were found to be more likely to show such damage than diurnally active birds, suggesting that they may frequently be involved in collisions (Goodman and Glynn 1988).

However, capturing prey at night, guided either by sound cues alone or by only gross spatial details from the visual system, is not sufficient in itself to account for the nocturnal habits of owls. It cannot, for example, account for general mobility within the bird's environment, whether to or from a hunting perch, returning to a perch after catching prey, or more general mobility concerned with nocturnal breeding behaviour (attendance at the nest, bringing food for the mate or young, etc) all of which must take place at low light levels. The problem of how the bird which is hunting by auditory cues alone can gauge the distance to, and identify, a possible prey item, must also be accounted for.

NOCTURNALITY AND NATURAL HISTORY: THE NOCTURNAL SYNDROME

To provide answers to these problems it is necessary to refer to various aspects of the natural history and behaviour of owls, as summarised in Chapter 7. It was noted that not all owls are equally nocturnal and that differences in nocturnality appeared to be correlated with (i) differences in the habitat types which were used for foraging and breeding, (ii) the dietary spectrum of the birds, and (iii) the degree to which the birds were territorial and sedentary. A strictly "nocturnal syndrome" was described, which linked a preference for a more continuous woodland habitat type with a highly sedentary life style. It was also argued that these owl species which are more flexible or variable in their degree of nocturnality, appear to forage over more open habitat types and tend not to use continuous woodlands for breeding. Although such birds are territorial in their breeding habits, they do not necessarily require continuous occupation of the same territory throughout their lives to ensure their survival.

The link between different aspects of these nocturnal life-styles becomes explicable in the light of the different limitations on the visual performance of owls in open and closed habitat types. The key lies in viewing the highly sedentary life style as essential to permit mobility within the darker and more spatially complex environment beneath a woodland canopy.

VISUAL PERCEPTION AND MINIMAL SENSORY INFORMATION

The influence of knowledge on seeing has been recognised since the middle of the 19th century when von Helmholtz first formulated the idea that much, if not all, perception in animals and man is dependent upon cognition (acquired knowledge) as well as immediate sensory input (see Gordon 1989 for an historical review of these ideas). Today the debate centres not on whether cognition makes any contribution to perception, but on the size of this contribution (see Ullman 1980 for an introduction to the "Direct" versus "Indirect" Perception debate). In the case of humans there are now many examples of experimental studies of visual perception which seem to support this idea, for example, Gregory (1974), Frisby (1979), Humphreys and Bruce (1988), and Gordon (1989). As well as many studies of the role of cognition in visual perception under controlled experimental conditions, there are also examples of investigations of more day-to-day kinds of phenomena which illustrate these ideas. These include such phenomena as the significant increase in the speed with which camouflaged, or hidden, objects can be detected or recognised as a result of practice or knowledge. The acquisition of detection and recognition skills by field ornithologists is but one example of how cognition (in this case the knowledge of which birds are likely to be present, aided by previous practised searching) facilitates the perception of a bird within the field of view. Other examples which are especially relevant concerning the mobility of owls at night, are studies of human mobility under reduced visual input. A particularly relevant example is provided by studies of car-driving, especially at night.

It has been shown that, when driving at night, people frequently, some-times habitually, drive beyond their "perceptual limit", i.e. they drive in a manner which relies on information which is not immediately available via their visual system (Hills 1980). Accidents are usually avoided, however, because the available visual information (e.g. detection of road markings and signs, and lights on other vehicles) is supplemented by experience of the nature of roads and traffic, and specific knowledge of the local road layout. It might seem that when driving at night the driver has full information about the road ahead, but in fact such information is really very limited, and the driver simply interprets this meagre information in a useful way. The driver assumes that the road will continue much as before, unless specific signals or signs suggest otherwise. Indeed, much modern road engineering is concerned with making roads as predictable as possible, with standard radius curves, constant road widths and marking conventions, so that only minimal cues are needed to drive safely.

That drivers are often beyond their perceptual limit is indicated by what happens when there is an unpredictable obstacle in the road, such as a pedestrian, stray animal, road works or broken-down car. At such times it is quite likely that an accident will happen, because the driver is travelling too fast to detect and correctly interpret what little information is available on the presence of the obstacle. On motorways and other types of fast main roads, it is widely acknowledged that drivers at night are travelling well beyond their perceptual limits. This is the main reason why road works and other obstructions have to be so well indicated; it often seems excessively so.

What this example shows is that mobility at night, even by man in common situations, is often achieved without guidance by the visual details that we accept, or might believe, are commonly available during the day. However, casual observation would not immediately suggest that a driver, travelling at speed along a stretch of road at night, is being guided by less information than when driving by day. Approximately equal performance at night to that by day is possible, because the reduced cues available can be correctly interpreted through a knowledge of roads in general and specific knowlede of local topography. To find out what those cues are, and the experience or knowledge necessary for their correct interpretation, requires considerable ingenuity (Hills 1980).

Car driving at night may not seem a nocturnal activity which is particularly relevant to mobility within a wood. However, it does seem likely that the general principles concerning the role of cognition in the correct interpretation of somewhat minimal cues, are of wide application. Indeed it is possible to think of many specific instances of nocturnal mobility in people where such principles might apply; for example, moving about one's home in semi-darkness guided only by minimal cues (the bright reflection from a door handle, the chink of light from a curtain, etc), or, perhaps, the local poacher hunting within his own regular patch of woodlands and not straying into other areas. The traditional poacher was a local man, who could only be caught at night by the local gamekeeper. It is also pertinent to note that approximately 80% of registered blind people do have some sight, and that they achieve competent locomotion, at least in their domestic environments, by a combination of reduced visual input and specific knowledge of their environment.

A SEDENTARY LIFE-STYLE AND NOCTURNAL MOBILITY

Because man and the owls are equally restricted at night in their main sensory capacities, a similar analysis might also apply to the nocturnal behaviour of

owls. Thus, it may be proposed that knowledge both of the general character-istics of the environment and specific details of the local topography are important for guiding an owl's behaviour under nocturnal conditions when visual cues are minimal.

Furthermore, it can be argued that a knowledge of local topography is more important when the conditions in which the owl is usually active become more exacting. For example, under a woodland canopy, compared with more open habitat types. In other words, the nocturnal habits of owls must depend upon behavioural adaptations which permit the accumulation of such knowl-edge, as well as sensory adaptations which maximise the chances that cues will be detected. Therefore, it can be proposed that the key factors which permit nocturnal mobility in the most exacting environments are strict territoriality and a sedentary life-style.

Aspects of the natural history of nocturnal woodland owls (such as the Tawny and Ural Owls), thus become explicable in the light of the sensory limitations and the cognitive aspects of nocturnal mobility, discussed above. Thus, their high degree of territoriality may be seen as essential to permit prey capture and general mobility when light levels limit the visual guidance of flight and other behaviour. The individual bird must be familiar with hunting perches, and the flight paths from them to the ground, not only to locate prey by hearing, but also to estimate the distance to the prey and then to return safely to a perch afterwards.

The owl, to be mobile, must also know flight paths through its habitat, so that it can avoid small obstacles and be guided only by the more gross spatial cues available at lower levels of illumination. To stray beyond its territory is of no advantage, because specific knowledge of local topography and the regularly used perches are essential for prey capture and movement under restricted sensory input.

This has two consequences, both of which have been demonstrated in the Tawny Owl (see Chapter 7). First, territorial boundaries are likely to remain fixed for long periods, and if an adjacent territory becomes vacant a neighbour is unlikely to expand into it. Secondly, when faced with a shortage of prey, the bird is not at liberty to forage further afield but must stay within its territory and accept a wide diversity of prey items, whether of optimal size or not.

Thus the highly sedentary life-style of nocturnal owls in woodland can be seen as essential if they are to forage and survive throughout the year. Residence within the same territory becomes essential for survival, not just for breeding success as is the case in most bird species.

It can similarly be argued that nocturnal owls which inhabit, or forage over, more open habitat are less restricted sensorily than woodland species, and are consequently less restricted in their life styles. As was shown in Figure 8.3, in open habitat, light levels, even on overcast moonless nights, are always likely to be above an owl's absolute visual threshold. Also, objects can still be seen in silhouette against the overcast. In addition, the surroundings in which the owl flies are likely to be spatially simple, and will therefore present fewer problems, compared to flight beneath a woodland canopy. It would not necessarily be disadvantageous for an owl, using this type of open habitat, to remain on its territory throughout the year since, on occasion, light levels will approach the bird's absolute threshold. Also, these species commonly use tree sites for nesting, and the approaches to them would still present perceptual problems. However, it seems likely that these birds could be mobile, at least over suitable open, foraging areas, without the need for specific topographical knowledge. Thus, nocturnal owls which use more open habitat types for foraging are likely to be less restricted to a specific locality, and can travel more widely outside the breeding season, perhaps in response to a local shortage of prey. For these owls, then, a territory may be necessary for breeding, but a sedentary life style is not essential for survival throughout the year, unlike the nocturnal owls in more continuous woodland habitats.

TERRITORIALITY, SPATIAL MEMORY AND NOCTURNALITY

The above discussion has proposed that strict territoriality and a sedentary life style are an essential component of the wholly nocturnal, woodland owl species. A territory is essential in these species not only to provide food for annual survival, but also to ensure that the bird can hunt that food under the full range of nocturnal conditions throughout the whole year. In times of shortage, food may be readily available in another area nearby, but, because its topography is unfamiliar, the owl is not likely to be very successful in catching that prey. Indeed, there is good evidence, for example in the Tawny and Ural Owls, that possession of a territory is essential for an owl's survival, not just for its breeding success. Also it is known that in some years the abundance of prey may decline such that successful breeding is not possible, yet the pair will remain on their territory rather than search further afield for prey, either to support themselves or their young (Southern 1970, Wendland 1984).

Thus, territoriality in these nocturnal woodland owls takes on a function additional to the more familiar one of providing food resources or nest sites (Wilson 1975; Davies 1978, 1980; Lundberg 1979; Wardough 1984) or reducing the probability of nest predation (Perrins 1979). Hinde (1956, p. 349) in an earlier review proposed that "familiarity with food sources and refuges from predators", could be regarded as one of the important functions of avian territorial behaviour. However, most recent work on territoriality has tended to favour analyses using metaphors drawn from economics in which emphasis has been upon the costs and benefits involved in the defence of resources within the territory.

Defence of a resource, and balancing the costs and benefits of its defence, implies that some commodity within the territory has a value which can be appropriated by other individuals, or can be readily given up in favour of more abundant resources elsewhere. However, topographical knowledge of a territory has value only to the resident. It cannot be transferred to another individual, nor can it be transferred by the resident to another site. Also, topographical knowledge, and hence the value of holding a particular terri-tory, is likely to increase with length of residence, much as an investment increases its value over time. Such benefits of territorial ownership in these nocturnal birds are somewhat intangible and difficult to quantify compared with measures of the resources, especially potential food supply or number of suitable nest sites, within the territory. However, some quantification is possible. Thus, Southern and Lowe (1968) have argued that continued occupancy of a territory by Tawny Owls results in greater hunting skill, and that this is reflected in an increase in the probability of successful breeding with the length of time that a pair has occupied a territory (Southern 1970).

It might be thought that familiarity both with the general topography of a complex woodland habitat (typically covering many hectares) and a more detailed knowledge of specific sites within it, would be a task that an owl could not achieve. Although there is no specific information on the ability of owls to learn spatial relationships, or on their long-term memory, data for other bird species has shown that memory for abstract visual information and for spatial relationships over long time periods can be most impressive (Macphail 1986).

For example, pigeons were taught to discriminate between 160 pairs of photographic slides showing a random selection of scenes (Vaughan and Green 1984). When tested for their retention of these discriminations 2 years later, it was found that practically all were remembered correctly.

Evidence that birds can remember an impressive number of spatial relation-ships comes from studies of food storage by Clark's Nutcracker *Nucifraga columbiana*. This bird, a North American member of the Corvidae, caches seeds from the pinon pine in autumn and recovers them over a period of months during the following winter and spring. Each bird may store up to 20,000–30,000 seeds in all, in some 5,000–8,000 caches. In order to survive, it appears that the nutcracker must find at least 2,500 caches. There is evidence from field studies (e.g. Tomback 1980; Vander Wall 1982) that memory is involved in cache recovery, rather than that the bird investigates sites where food might have been stored.

Kamil and Balda (1985) have shown that Clark's Nutcrackers can recover seeds successfully even when the location of the caches has been predetermined by the experimenters, thus increasing the likelihood that the birds were, indeed, using spatial memory to recover the seeds many weeks after they had been hidden. Even in the Marsh Tit *Parus palustris,* it has been shown that the birds can remember the location of approximately 70 sites where food was cached between 3–19 days earlier (Cowie *et al.* 1981).

It has also been shown that other passerine species are aware of the spatial layout of their foraging areas by studies of "neophobia" (Greenberg 1983, 1987). In these studies it was clear that warbler species (Parulinae) had sufficient knowledge of their foraging area to detect immediately the presence of novel objects placed there by the investigator.

It has also been demonstrated that various birds species, including Garden Warbler *Sylvia borin* (Biebach *et al* 1989), hummingbirds (Cole *et al* 1982), Kestrel *Falco tinnunculus* (Rijnsdoorp *et al* 1981) and Oystercatcher *Haematopus ostralegus* (Daan and Koene 1981), apparently learn both the place and time where food regularly occurs, thus suggesting that a highly developed spatial memory may be commonly used by many birds when foraging (Olton *et al* 1981).

It is also worth noting that Baker (1984) has proposed that when completing their migrations between Europe and Africa, birds do so by building up a series of cognitive maps. These maps form a mosaic, linking areas whose topography is known in detail (where suitable foraging sites may be found), and routes between them. On the return and subsequent migrations the birds may consult their cognitive maps to determine their route and then to recognise their breeding area on arrival. In addition, it is widely held that Homing Pigeons find their way to their lofts by reference to a sophisticated map-sense, which may be based on a variety of topographical, magnetic and olfactory cues (Gould 1982).

Thus, it would seem that learning spatial relationships, and retaining that information over relatively long periods, may be important in birds of widely differing life styles. In the case of the nocturnal, sedentary owls, it may be that their cognitive map covers just a few hectares but is highly detailed. On the other hand, migrant birds and those which home to nest sites, have much less detailed maps, but ones which may span a large slice of the globe.

CHAPTER 10

Further light on night birds?

This book has considered a wide range of examples of nocturnal activity in birds. Its coverage of individual species which are active at night is certainly not comprehensive. Many further examples exist of observations of birds engaging in occasional nocturnal or crepuscular activities. These may be as diverse as a European Robin *Erithacus rubecula* singing beneath a suburban street light, the leisurely "filter feeding" of Greater Flamingos *Phoenicopterus ruber* in a shallow Mediterranean lagoon, the more stealthy feeding of Night Herons *Nycticorax nycticorax* upon frogs as they begin to call at dusk, or the more frantic feeding of Crab Plovers *Dromas ardeola* as they gorge themselves on young crabs emerging upon a moonlit ocean shore. The reader, from their own experience and from published sources, can probably add many more examples of occasional crepuscular and nocturnal activities, all of which deserve further discussion and more systematic investigation.

It does seem likely, however, that all important types of occasional and regular nocturnal activity have been considered in the text. For each instance the aim has been to examine the type of behaviour involved, the particular circumstances in which it occurs and to consider any available information on the sensory capacities which may permit such nocturnal activity. It has been seen that instances of nocturnal behaviour present problems of considerable complexity all of which require more data before the behaviour is fully understood. Perhaps the main function of the book has been to clarify some of the problems and call to question some assumptions.

180

A few general points have emerged from these discussions and it is worth considering them briefly here.

SENSORY CAPACITIES AND NOCTURNAL FORAGING

To understand nocturnal behaviour in birds it is essential to appreciate that there are unlikely to be any simple sensory explanations. That is, there are no "super senses" which might account for nocturnal behaviour in any bird. It would seem that there are ultimate limits to the sensitivity, both of vision and hearing, in all terrestrial vertebrates, and that while the visual and auditory systems of nocturnal birds may be able to achieve those limits, they cannot surpass them.

Nocturnal foraging depends upon an intimate set of relationships between the habitat type in which it is performed, sensory capacities and specific behaviours. The actual detection and capture of prey at night can usually occur without visual guidance. However, the sensory cues which supplement vision may differ considerably; tactile and taste sensitivity in the bill in the case of the waders, waterfowl and skimmers, and possibly in the nightjars; olfaction in petrels, kiwis, and perhaps in the Oilbird; hearing in the owls, and possibly in the frogmouths and potoos, thick-knees and coursers.

The range of food items taken by nocturnal foragers is usually restricted to two main types. (1) Items which occur at a low density but advertise their presence by giving out noise or a strong odour, and (2) those which occur at high density and can be detected using touch and taste sensitivity and random searching.

FLIGHT AT NIGHT

Where flight occurs at night it nearly always takes place (with the exception of certain species of owls, frogmouths and potoos) in open habitats, often well clear of vegetation or other obstructions. It seems that this is for two reasons. First, in all species, visual capacities are never sufficient to permit the avoidance of small natural obstacles under many night-time conditions and, secondly, that light levels will always be significantly higher in open habitats compared with those under a vegetation canopy. Even those owls which are active at night in more spatially complex habitats, are probably unable to detect fine spatial detail at many nocturnal light levels.

It does seem, however, that many birds are willing to fly completely or almost blind, guided only by very coarse spatial cues. This even applies to the echolocatory birds which fly in the totally dark interior of caves. However, birds which habitually fly at night, guided by coarse spatial detail, tend to have wings whose structure permits slow flight or even momentary hovering (e.g. owls, nightjars, Oilbird). Flight at night, both in regularly and occasionally nocturnal birds, is not without hazard. Birds may become perceptually confused, as in the case of nocturnally migrating passerines attracted to lights or a surprised owl crashing into branches.

GENERAL LESSONS FROM THE OWLS

It has been argued that the limits on visual and auditory sensitivity in owls are not unique to these birds. Equally, the relationship between these limits on sensitivity and the actual sensory problems presented by the night environment are likely to have universal application. Hence it seems likely that the "nocturnal syndromes", linking various degrees of nocturnality, territoriality and habitat types, described in Chapter 7, should apply generally to owl species and also to other nocturnal birds, including many species among the Caprimulgiformes, especially the frogmouths and potoos. The arguments put forward, here, certainly show that to achieve a truly nocturnal life style involves exacting behavioural adaptations, and this must go some way to explain why strict nocturnality is so rare among birds. Furthermore, some of the points raised here, concerning limits on the visual guidance of behaviour under nocturnal conditions, are likely to apply also to nocturnal activity in other animal groups, including mammals. It would clearly be worth considering the extent to which these animals are also dependent upon familiarity with their surroundings in order to achieve nocturnal mobility throughout the year.

FURTHER PROBLEMS

The scope for further investigations into the occurrence of nocturnal behaviour in birds, and its associated behavioural and sensory adaptations, is vast. Many questions have been raised but there are few definitive answers. Some questions concern the sensory basis of nocturnal foraging, for example, in nightjars, Oilbird, frogmouths, diving ducks, thick-knees and other waders. Many questions can be asked about the relationships between light levels and the occurrence of occasional nocturnal behaviour and foraging efficiency in a wide range of species. Also, the particular circumstances under which certain birds choose to migrate at night is clearly worthy of investigation.

However, as was seen when attempting to unravel the problems of nocturnality in owls, information from many different sources is necessary. Field studies of ecology and behaviour, laboratory experiments on prey-catching and foraging techniques, investigations of sensory capacity, sensory physiology and anatomy, are all necessary if a fuller understanding is to be achieved.

It is to be hoped that this book will have aroused sufficient curiosity to encourage the more determined day-time ornithologists to take a closer look at what birds do at night, and to consider how they might do it.

Appendix 1

Examples of the reports of the normal pattern of diurnal and nocturnal migratory activity of certain bird species which pass through bird observatories sited around the coast of the British Isles. Observers were asked to assign species to one of five categories as follows: D, exclusively diurnal; D(N), most individuals diurnal but occasional nocturnal activity; ND, equally nocturnal and diurnal; N(D), most individuals nocturnal but occasional diurnal activity; N, exclusively nocturnal migrant. Only those species for which responses from at least three observatories were received were included in the analysis. Based upon these category judgements each species was allotted to one of the five categories, the final category. In this, a species was allotted to an exclusively diurnal or nocturnal category only if all reports for that species agreed, otherwise the species was placed in either of the other three categories which involved nocturnal and diurnal migration to different degrees.

The names and geographical positions of the observatories are as follows: 1. Copeland, small island 4 km off the east coast of Northern Ireland, 54° 41' N 5° 32' W. 2. Calf of Man, small island 1 km off the south west tip of the Isle of Man, 54° 03' N 4° 49' W. 3. Walney Island on the north-west side of Morecambe Bay, Cumbria, 54° 05' N 3° 15' W. 4. Bardsey, small island 3 km off the Lleyn Peninsula, North Wales, 52° 46' N 4° 48' W. 5. Dungeness, large shingle peninsula near Lydd, Kent, 50° 55' N 0° 57' E. 6. Spurn, northern end of narrow coastal peninsula, North Humberside 53° 35' N 0° 06' E. 7. Fair Isle, small island midway between Orkney and Shetland, North Sea, 59° 32' N 1° 37' W.

[The English names of the bird species refer to those used in Hudson R. (ed) (1987) A species list of British and Irish birds. (Third edition). British Trust for Ornithology, Tring.]

	Observatories							Final category
Species	1	2	3	4	5	6	7	
Charadriiformes								
Oystercatcher	ND	ND	N(D)	D(N)	D(N)	D(N)	ND	ND
Ringed Plover		D	D(N)	D(N)	ND	D		ND
Golden Plover		D	N(D)	D(N)		D(N)	ND	ND
Grey Plover			N(D)			D(N)	ND	ND
Lapwing	D		D(N)	D	D(N)	D(N)	ND	D(N)
Knot			N(D)			D(N)	ND	ND
Sanderling			N(D)	N(D)		D(N)	ND	ND
Dunlin	N	D	N(D)	ND	ND	D(N)	ND	ND
Ruff			N(D)			N(D)	ND	ND
Snipe		ND	N(D)	ND	N(D)	D	ND	N(D)
Woodcock		N	N	N(D)		N	ND	N(D)
Bar-tailed Godwit	D	D	N(D)	D(N)	D(N)	D(N)	ND	ND
Whimbrel	D	ND	D(N)	D(N)	D(N)	D(N)	ND	D(N)
Curlew	D	D	D(N)	D(N)		D(N)	ND	D(N)
Redshank		D(N)	N(D)	ND	ND	D	ND	ND
Greenshank		D(N)	N(D)		ND	N(D)	ND	ND
Green Sandpiper			N(D)		ND	N(D)	ND	ND
Wood Sandpiper			N(D)	ND		N(D)	ND	ND
Common Sandpiper		N(D)	N(D)	N(D)	N	N	ND	N(D)
Turnstone	D	D	N(D)	D(N)		D(N)	ND	D(N)

Sylviidae								
Grasshopper Warbler	N	N	N(D)	N	ND	N	ND	N(D)
Sedge Warbler	N	N	N(D)	N	ND	N	ND	N(D)
Reed Warbler					ND	N	ND	ND
Icterine Warbler					ND	N	ND	ND
Barred Warbler	N				ND	N		ND
Lesser Whitethroat					ND	N		ND
Whitethroat	N	N	N(D)	N	ND	N	ND	N(D)
Garden Warbler	N	N	N(D)	N	ND	N	ND	N(D)
Blackcap	N	N	N(D)	N	ND	N	ND	N(D)
Yellow-browed Warbler	N				ND	N	ND	N(D)
Wood Warbler					ND	N		ND
Chiffchaff	N	N(D)	N(D)	N	ND	N	ND	N(D)
Willow Warbler	N	N(D)	N(D)	N	ND	N	ND	N(D)
Goldcrest	N		N(D)	N	ND	N	ND	N(D)
Firecrest					ND	N	ND	ND
Fringillidae								
Chaffinch	ND	D(N)	D	D	D	ND		D(N)
Brambling	ND	D	D	D	D	ND		D(N)
Greenfinch	D	D	D	D	D	D		D
Goldfinch	D	D	D	D	D	D		D
Siskin	D	D	D	D	D	ND		D(N)
Linnet		D	D	D	D	D		D
Twite		D	D	D	D	D		D
Redpoll	D	D	D	D	D	ND		D(N)

Appendix 2

The pattern of diurnal and nocturnal migratory activity of 147 bird species which pass through bird observatories around the coast of the British Isles. Birds were assigned to one of five categories according to the reports from individual bird observatories as presented in Table 4.1. D, exclusively diurnal; D(N) most individuals diurnal but occasional nocturnal activity; ND, equally nocturnal and diurnal; N(D), most individuals nocturnal but occasional diurnal activity.

D	D(N)	ND	N(D)
Gaviiformes			
Red-throated Diver			
Black-throated Diver			
Great Northern Diver			
Podicipediformes			Little Grebe
	Grt. Crested Grebe		
Procellariiformes			
Fulmar			
Manx Shearwater		Leach's Petrel	
Pelecaniformes			
Gannet			
Cormorant			

Ciconiiformes

Grey Heron

Anseriformes

Bewick's Swan
Whooper Swan

Pink-footed Goose
Greylag Goose
Barnacle Goose
Brent Goose
Shelduck

Wigeon
Teal
Mallard
Tufted Duck

Eider
Long-tailed Duck
Common Scoter
Velvet Scoter
Goldeneye
Red-breasted Merganser

Accipitriformes

Marsh Harrier
Hen Harrier

Sparrowhawk
Buzzard

Rough-legged Buzzard

Osprey

(continued)

Appendix 2—continued

D	D(N)	ND	N(D)
Falconiformes			
Kestrel			
Merlin			
Peregrine	Hobby		
Gruiformes			Water Rail
Charadriiformes		Oystercatcher	
		Ringed Plover	
		Golden Plover	
		Grey Plover	
	Lapwing	Knot	
		Sanderling	
		Dunlin	
		Ruff	
		Snipe	
	Whimbrel	Bar-tailed Godwit	Woodcock
	Curlew		

Pomarine Skua
Arctic Skua
Great Skua

Turnstone

Redshank
Greenshank
Green Sandpiper
Wood Sandpiper

Common Sandpiper

Black-headed Gull
Common Gull

Lesser Black-backed Gull
Herring Gull

Great Black-backed Gull
Kittiwake
Sandwich Tern
Common Tern
Arctic Tern
Little Tern
Black Tern
Guillemot
Razorbill
Puffin

Columbiformes

Woodpigeon
Collared Dove
Turtle Dove

(*continued*)

Appendix 2—*continued*

D	D(N)	ND	N(D)
Cuculiformes	Cuckoo		
Strigiformes		Long-eared Owl Short-eared Owl	
Apodiformes	Swift		
Piciformes		Wryneck Great Spotted W'pecker	
Passeriformes *Alaudidae*	Skylark		
Hirundinidae Sand Martin Swallow	House Martin		
Motacillidae	Tree pipit Meadow Pipit		
Rock/Water Pipit	Yellow Wagtail Grey Wagtail Pied/White Wagtail		

Troglodytidae

Wren

Prunellidae

Dunnock

Turdidae

Robin
Black Redstart
Redstart
Whinchat
Stonechat
Wheatear
Ring Ousel
Blackbird
Fieldfare
Song Thrush
Redwing
Mistle Thrush

Sylviidae

Grasshopper Warbler
Sedge Warbler
Reed Warbler
Icterine Warbler
Barred Warbler
Lesser Whitethroat
Whitethroat
Garden Warbler
Blackcap
Yellow-browed W.

(continued)

D	D(N)	ND	N(D)
Passeriformes, Sylviidae—continued			
		Wood Warbler	Chiffchaff Willow Warbler Goldcrest
		Firecrest	
Muscicapidae			
			Spotted Flycatcher Pied Flycatcher
Paridae Blue Tit Great Tit			
Oriolidae		Golden Oriole	
Lanidae		Red-backed Shrike	
Corvidae Magpie Jackdaw Rook Carrion/Hooded Crow			

Sturnidae
Starling

Passeridae
Tree Sparrow
Fringillidae
Chaffinch
Brambling
Greenfinch
Goldfinch
Siskin
Linnet
Twite
Redpoll

Emberizidae
Lapland Bunting
Snow Bunting
Reed bunting

References

Able, K. P. 1974. Environmental influences on the orientation of free-flying nocturnal bird migrants. *Anim. Behav.* 22: 224–238.

Able, K. P. 1980. Mechanisms of orientation, navigation and homing. Pp. 283–373 in *Animal Migration, Orientation and Navigation* (ed. S. A. Gauthreaux Jr). Academic Press, New York.

Able, K. P. 1982. Field studies of avian nocturnal migratory orientation. 1. Interaction of sun, wind and stars as directional cues. *Anim. Behav.* 30: 761–767.

Able, K. P. & Bingman, V. P. 1987. The development of orientation and navigation behaviour in birds. *Q. Rev. Biol.* 62: 1–29.

Able, K. P. & Cherry, J. D. 1986. Mechanism of dusk orientation in white-throated sparrows (*Zonotrichia albicollis*): clock-shift experiments. *J. Comp. Physiol.* 159: 107–113.

Able, K. P. & Gauthreaux, S. A. 1975. Quantification of nocturnal passerine migration with a portable ceilometer. *Condor* 77: 92–96.

Albone, E. S. 1984. *Mammalian Semiochemistry*. Wiley.

Aldrich, J. W., Graber, R. R., Munro, D. A., Wallace, G. J., West, G. C. & Gahalane, V. H. 1966. Mortality at ceilometers and towers. *Auk* 83: 465–467.

194

Alerstam, T. 1985. Radar. Pp. 492–494 in *A Dictionary of Birds* (eds B. Campbell & E. Lack). T. & A. D. Poyser.

Alerstam, T. & Pettersson, S.-G. 1976. Do birds use waves for orientation when migrating across the sea? *Nature* (Lond.) 259: 205–207.

Amlaner, C. J. & Ball, N. J. 1983. A synthesis of sleep in wild birds. *Behaviour* 87: 85–119.

Ammons, C. H., Worchel, P. & Dallenbach, K. M. 1953. "Facial vision": the perception of obstacles out of doors by blindfolded and blindfold-deafened subjects. *Am. J. Psychol.* 66: 519–553.

Armstrong, E. A. 1958. *The Folklore of Birds.* Collins.

Armstrong, E. A. 1963. *A Study of Bird Song.* Oxford University Press.

Aschoff, J. 1967. Circadian rhythms in birds. *Proc. XIV Int. Orn. Congr.* 81–105.

Aschoff, J., Gwinner, E., Kureck, A. & Muller, K. Diel rhythms of chaffinches (*Fringilla coelebs*), tree schrews (*Tupaia glis*) and hamsters (*Mesocricetus auratus*) as a function of season at the Arctic Circle. *Oikos* Suppl. 13: 91–100.

Astrom, G. 1976. Environmental influences on daily song activity of the Reed Bunting (*Emberiza schoeniclus (L).)* *Zoon* Suppl. 2: 1–82.

(The) Astronomical Almanac. 1988. H.M.S.O.

Aubert, H. 1865. *Physiologie der Netzhant.* Morgenstern, Breslau.

Bacon, P. J. 1985. Roosting. Pp. 517–519 in *A Dictionary of Birds* (eds B. Campbell & E. Lack). T. & A. D. Poyser.

Baker, R. R. 1978. *The Evolutionary Ecology of Animal Migration.* Hodder & Stoughton.

Baker, R. R. 1984. *Bird Navigation: the Solution of a Mystery?* Hodder & Stoughton.

Baldwin, S. P. & Kendeigh, S. C. 1938. Variation in the weight of birds. *Auk* 55: 416–467.

Bang, B. G. 1966. The olfactory apparatus of tubenosed birds *(Procellariiformes). Acta Anat.* 65: 391–415.

Bang, B. G. 1971. Functional anatomy of the olfactory system in 23 orders of birds. *Acta Anat.* 79: Suppl. 1–76.

Bang, B. G. and Cobb, S. 1968. The size of the olfactory bulb in 108 species of bird. *Auk* 85: 55–61.

Bang, B. G. & Wenzel, B. M. 1985. Nasal cavity and olfactory system. Pp. 195–225 in *Form and Function in Birds,* vol. 3 (eds A. S. King & J. McLelland). Academic Press.

Barlow, H. B. 1962. Measurements of the quantum efficiency of discrimination in human scotopic vision. *J. Physiol.* 160: 169–188.

Barlow, H. B. 1972. Dark and light adaptation: psychophysics. Pp. 1–28 in *Handbook of Sensory Physiology,* vol. VII/4 (eds D. Jameson & L. M. Hurvich). Springer-Verlag, Berlin.

Barlow, H. B. 1981. Critical limiting factors in the design of the eye and visual cortex. *Proc. R. Soc. Lond. B* 212: 1–34.

Barrett, R., Maderson, P. F. A. & Meszler, R. M. 1970. The pit organs of snakes. Pp. 277–304 in *Biology of the Reptilia* (eds C. Gans & T. S. Parsons). Academic Press.

Bartlett, N. R. 1965. Dark adaptation and light adaptation. Pp. 185–207 in *Vision and Visual Perception* (ed. C. H. Graham). Wiley, New York.

Baumgardt, E. 1972. Threshold quantal problems. Pp. 29–55 in *Handbook of Sensory Physiology,* vol. VII/4 (eds D. Jameson & L. M. Hurvich). Springer-Verlag, Berlin.

Becking, J. H. 1971. The breeding of *Collacalia gigas. Ibis* 113: 330–334.

Beecher, W. J. 1978. Feeding adaptations and evolution in the starlings. *Bull. Chicago Acad. Sci.* 11: 269–298.

Bellrose, F. C. 1967. Radar in orientation research. *Proc. XIV Int. Orn. Congr.* 281–309.

Bellrose, F. C. & Graber, R. R. 1963. A radar study of the flight direction of nocturnal migrants. *Proc. XIII Int. Orn Congr.* 363–389.

Berkhoudt, H. 1977. Taste buds in the bill of the Mallard (*Anas platyrhynchos L.*): their morphology, distribution and functional significance. *Netherl. J. Zool.* 27: 310–331.

Berkhoudt, H. 1980. The morphology and distribution of cutaneous mechanoreceptors (Herbst and Gandry corpuscles) in bill and tongue of the Mallard (*Anas platyrhynchos L.*) *Netherl. J. Zool.* 30: 1–34.

Berkhoudt, H. 1985. Structure and function of avian taste receptors. Pp. 463–496 in *Form and Function in Birds,* vol. 3 (eds A. S. King & J. McLelland). Academic Press.

Berry, R. 1979. Nightjar habitats and breeding in East Anglia. *British Birds* 72: 207–218.

Biebach, H., Friedrich, W. & Heine, G. 1986. Interaction of bodymass, fat, foraging and stopover period in trans-Sahara migrating passerine birds. *Oecologia* 69: 370–379.

Biebach, H., Gordijn, M. & Krebs, J. R. 1989. Time-and-place learning by garden warblers, *Silvia borin. Anim. Behav.* 37: 353–360.

Bingman, V. P., Able, K. P. & Kerlinger, P. 1982. Wind drift, compensation, and the use of landmarks by nocturnal bird migrants. *Anim. Behav.* 30: 49–53.

Blase, B. 1971. Zum Beginn und Ende der taglichen Aktivitat de Goldammer. *Falke* 18: 228–241.

Blough, D. S. 1955. Method for tracing dark adaptation in the pigeon. *Science* 121: 703–704.

Blough, D. S. 1956. Dark adaptation in the pigeon. *J. Comp. Physiol. Psychol.* 49: 425–430.

Blough, P. M. 1971. The visual acuity of the pigeon for distant targets. *J. Exp. Anal. Behav.* 15: 57–68.

Bolze, G. 1968. Anordnung und Bau der Herbstschen Korperchen in Limicolenschnabeln im Zusammenhang mit der Nahrungsfindung. *Zool. Anz.* 181: 313–355.

Bond, D. S. & Henderson, F. P. 1963. *The Conquest of Darkness.* AD346297. Defense Documentation Centre, Alexandria, Va.

Bourne, W. R. P. 1982. The midnight descent, dawn ascent and re-orientation of land birds migrating across the North Sea in autumn. *Ibis* 122: 536–540.

Bourne, W. R. P. 1985. Petrel. Pp. 451–456 in *A Dictionary of Birds* (eds B. Campbell & E. Lack). T. & A. D. Poyser.

Bowmaker, J. K. 1977. The visual pigments, oil droplets and spectral sensitivity of the pigeon. *Vision Res.* 17: 1129–1138.

Bowmaker, J. K. & Dartnall, H. J. A. 1980. Visual pigments of rods and cones in a human retina. *J. Physiol. (Lond.)* 298: 501–511.

Bowmaker, J. K. & Knowles, A. 1977. The visual pigments and oil droplets of the chicken retina. *Vision Res.* 17: 755–764.

Bowmaker, J. K. & Martin, G. R. 1978. Visual pigments and colour vision in a nocturnal bird, *Strix aluco* (Tawny Owl). *Vision Res.* 18: 1125–1130.

Bradbury, J. W. 1970. Target discrimination by the echolocating bat *Vampyrum spectrum. J. Exp. Zool.* 173: 23.

Brooke, M. de L. 1978a. A test for visual location of the burrow by Manx Shearwaters, *Puffinus puffinus. Ibis* 120: 347–349.

Brooke, M. de L. 1978b. Sexual differences in the voice and individual vocal recognition in the Manx Shearwater (*Puffinus puffinus*). *Anim. Behav.* 26: 622–629.

Brooke, M. de L. & Klages, N. 1986. Squid beaks regurgitated by Greyheaded and Yellownosed albatrosses, *Diomedea chrysostoma* and *D. chlororhynchos* at the Prince Edward Islands. *Ostrich* 57: 203–206.

Brooke, R. K. 1970. Taxonomic and evolutionary notes on the subfamilies, tribes, genera and subgenera of the swifts (Aves: Apodidae). *Durban Mus. Novit.* 9: 13–24.

Brown, J. L. & Mueller, C. G. 1965. Brightness discrimination and brightness contrast. Pp. 208–250 in *Vision and Visual Perception* (ed. C. H. Graham). Wiley, New York.

Brown, L. 1976a. *Birds of Prey: their Biology and Ecology.* David & Charles.

Brown, L. 1976b. *British Birds of Prey.* Collins.

Brown, L. & Amadon, D. 1968. *Eagles, Hawks and Falcons of the World.* David & Charles.

Bruderer, B. 1978. Effects of alpine topography and winds on migrating birds. Pp. 252–265 in *Animal Migration, Navigation and Homing* (eds K. Schmidt-Koenig and W. T. Keeton). Springer-Verlag, Berlin.

Bruderer, B. 1982. Do migrating birds fly along straight lines? Pp. 3–14 in *Avian Navigation* (eds F. Papi & H. G. Wallraff). Springer-Verlag, Berlin.

Buhler, P. 1970. Schademorphologie und Kiefermechanik der Caprimulgidae (Aves). *Z. Morph. Tiere* 66: 337–399.

Buhler, P. 1981. The functional anatomy of the avian jaw apparatus. In *Form and Function in Birds,* vol. 2 (eds A. S. King and J. McLelland). Academic Press.

Bullock, T. H. and Diecke, F. D. J. 1956. Properties of an infra-red receptor. *J. Physiol. (Lond.)* 134: 47–87.

Bunn, D. S., Warburton, A. B. & Wilson, R. D. S. 1982. *The Barn Owl.* T. & A. D. Poyser.

Burton, J. A. (ed.) 1984. *Owls of the World.* Peter Lowe.

Burton, P. J. K. 1974. *Feeding and Feeding Apparatus in Waders.* British Museum.

Catchpole, C. K. 1979. *Vocal Communication in Birds.* Edward Arnold.

Cemmick, D. & Veitch, D. 1988. *Kakapo Country.* Hodder & Stoughton.

Clark, L. & Mason, J. R. 1987. Olfactory discrimination of plant volatiles by the European Starling. *Anim. Behav.* 35: 227–235.

Clark, R. 1975. A field study of the Short-eared Owl, *Asio flammeus* (Pontoppidan) in North America. *Wildlife Monographs* 47: 1–67.

Clark, R., Smith, D. & Kelso, L. 1978. *Working Bibliography of Owls of the World.* Nat. Wildlife Fed. (Washington) Sci. Techn. Ser. 1: 1–319.

Clarke, M. R., Croxall, J. P. & Prince, P. A. 1981. Cephalopod remains in regurgitations of the Wandering Albatross *Diomedea exulans* at South Georgia. *Brit. Ant. Surv. Bull.* 54: 9–21.

Clarke, M. R. & Prince, P. A. 1981. Cephalopod remains in the regurgitations of Black-browed and Grey-headed Albatrosses at South Georgia. *Brit. Ant. Surv. Bull.* 54: 1–7.

Clunie, F. 1976. The Fiji peregrine (*Falco peregrinus*) in an urban-marine environment. *Notoris* 23: 8–28.

Cobb, S. 1960. A note on the size of the avian olfactory bulbs. *Epilepsia* 1: 394–402.

Cole, S., Hainsworth, F. R., Kamil, A. C., Mercier, T. & Wolf, L. L. 1982. Spatial learning as an adaptation in humming birds. *Science N.Y.* 217: 655–657.

Coleman, P. 1962. Failure to localise the source distance of an unfamiliar sound. *J. Acoust. Soc. Am.* 34: 345–346.

Cowie, R. J., Krebs, R. J. & Sherry, D. F. 1981. Food storing by marsh tits. *Anim. Behav.* 29: 1252–1259.

Cowles, G. S. 1967. The palate of the red-necked nightjar *Caprimulgus ruficollis* with a description of a new feature. *Ibis* 109: 260–265.

Cramp, S. (ed.) 1985. *The Birds of the Western Palearctic,* vol. IV. Oxford University Press.

Cramp, S. (ed.) 1988. *The Birds of the Western Palearctic,* vol. V. Oxford University Press.

Cramp, S. & Simmons, K. E. L. (eds) 1977. *The Birds of the Western Palearctic,* vol. I. Oxford University Press.

Cramp, S. & Simmons, K. E. L. (eds) 1980. *The Birds of the Western Palearctic,* vol. II. Oxford University Press.

Cramp, S. & Simmons, K. E. L. (eds) 1983. *The Birds of the Western Palearctic,* vol. III. Oxford University Press.

Croxall, J. P., Hill, H. J., Lidstone-Scott, R., O'Connell, M. J. & Prince, P. A. 1988. Food and feeding ecology of Wilson's Storm Petrel *Oceanites oceanicus* at South Georgia. *J. Zool. Lond.* 216: 83–102.

Curry-Lindahl, K. 1981. *Bird Migration in Africa,* vol. 2. Academic Press.

Cuthill, I. C. & Macdonald, W. A. 1990. Experimental manipulation of the dawn and dusk chorus in the blackbird *Turdus merula. Behav. Ecol. Sociobiol.* 26: 209–216.

Daan, S. & Koene, P. 1981. On the timing of foraging flights by oystercatchers *Haematopus ostralegus* on tidal mudflats. *Neth. J. Sea Res.* 15: 1–22.

Davies, N. B. 1978. Ecological questions about territorial behaviour. Pp. 317–350 in *Behavioural Ecology: an Evolutionary Approach* (eds J. R. Krebs & N. B. Davies). Oxford University Press.

Davies, N. B. 1980. The economics of territorial behaviour in birds. *Ardea* 68: 63–74.

Dawson, E. W. 1978. Kiwis. Pp. 25–26 in *Bird Families of the World* (ed. C. J. O. Harrison). Elsevier-Phaidon.

Dice, D. L. 1945. Minimum intensities of illumination under which owls can find dead prey by sight. *Am. Nat.* 79: 384–416.

Diercks, K. J. & Jeffress, L. A. 1962. Interaural phase and the absolute threshold for tone. *J. Acoust. Soc. Am.* 34: 981–984.

Dooling, R. J. 1982. Auditory perception in birds. Pp. 95–130 in *Acoustic Communication in Birds,* vol. 1 (eds D. Kroodsma & E. Miller). Academic Press.

Doring, G. 1920. Das erwachen der Vogelwelt im hoheren sachsischen Erzgebirge zu den verschiedenen Jahreszeiten. *Tharandt. forstl. Jb.* 71: 242–263.

Dorno, C. 1924. Reizphysiologische Studien uber den Gesang der Vogel im Hochgebirge. *Pflug. Arch. ges. Physiol.* 204: 645–659.

Dorst, J. 1961. *The Migration of Birds.* Heinemann.

Drury, W. H. & Nisbet, I. C. T. 1964. Radar studies of orientation of songbird migrants in south-eastern New England. *Bird Banding* 35: 69–119.

Dugan, P. J. 1981. The importance of nocturnal foraging in shorebirds: a consequence of increased vertebrate prey activity. Pp. 251–260 in *Feeding and Survival Strategies of Estuarine Organisms* (eds N. V. Jones & W. J. Wolff). Plenum Press, New York.

Durman, R. 1976. Bardsey. Pp. 29–46 in *Bird Observatories in Britain and Ireland* (ed. R. Durman). T. & A. D. Poyser.

Eastwood, E. 1967. *Radar Ornithology.* Methuen.

Elkins, N. 1983. *Weather and Bird Behaviour.* T. & A. D. Poyser.

Emlen, S. T. 1967. Migratory orientation in the indigo bunting *Passerina cyanea.* 1. The evidence for celestial cues. *Auk* 84: 309–312.

Emlen, S. T. 1975. Migration: orientation and navigation. Pp. 129–219 in *Avian Biology,* vol. 5 (eds D. S. Farner & J. R. King). Academic Press.

Emlen, S. T. & Demong, N. J. 1978. Orientation strategies used by free-flying bird migrants: a radar tracking study. Pp. 283–293 in *Animal Migration, Navigation and Homing* (eds K. Schmidt-Koenig and W. T. Keeton). Springer-Verlag, Berlin.

Emmerton, J. & Delius, J. D. 1980. Wave length discrimination in the "visible" and ultraviolet spectrum by pigeons. *J. Comp. Physiol.* 141: 47–52.

Engelmann, C. 1957. *So Leben Huhner, Tauber, Ganse.* Neumann Verlag, Radebeal.

Epple, G. 1986. Communication by chemical signals. In *Comparative Primate Biology*, vol. 2A (eds G. Mitchel and J. Erwin). Alan R. Liss, New York.

Erwin, R. M. 1977. Black Skimmer breeding ecology and behaviour. *Auk* 94: 709–717.

Evans, P. R. 1976. Energy balance and optimal foraging strategies in shorebirds: some implications for their distribution and movement in the non-breeding season. *Ardea* 64: 117–139.

Evans, P. R. 1985. Migration. Pp. 348–353 in *A Dictionary of Birds* (eds B. Campbell & E. Lack). T. & A. D. Poyser.

Evans, P. R. 1990. Strategies of migration in waders. In *Bird migration: The physiology and ecophysiology* (ed. E. Gwinner). Springer-Verlag, Berlin.

Fallet, M. 1962. Uber Bodenvogel und ihre terricolen Beutetiere: Technick der Nahrungssuche-Populationsdynamik. *Zool. Anz.* 168: 187–212.

Feare, C. J. 1984. *The Starling*. Oxford University Press.

Fedducia, A. 1985. Flightlessness. Pp. 223–224 in *A Dictionary of Birds* (eds B. Campbell & E. Lack). T. & A. D. Poyser.

Federer, C. A. & Tanner, C. B. 1966. Spectral distribution of light in a forest. *Ecology* 47: 555–560.

Fenton, M. B. Acuity of echolocation in *Collacalia hirundinacea* (Aves: Apodidae), with comments on the distribution of echolocating swiftlets and molossid bats. *Biotropica* 7: 1–7.

Fite, K. V. 1973. Anatomical and behavioural correlates of visual acuity in the Great Horned Owl. *Vision Res.* 13: 219–230.

Fleay, D. 1968. *Night Watchmen of Bush and Plain*. Jacaranda Press, Brisbane.

Forshaw, J. M. 1978. *Parrots of the World*. 2nd ed. David & Charles.

Frisby, J. 1979. *Seeing, Illusion, Brain and Mind*. Oxford University Press.

Gans, C. & Parsons, T. S. (eds) 1970. *Biology of the Reptilia*. Academic Press.

Gatehouse, R. W. & Shelton, B. R. 1978. Sound localisation in Bobwhite Quail (*Colinus virginianus*). *Behavioural Biol.* 22: 533–540.

Gauthreaux, S. A. 1978. Importance of daytime flights of nocturnal migrants: redetermined migration following displacement. Pp. 219–227 in *Animal Migration, Navigation and Homing* (eds K. Schmidt-Koenig & W. T. Keeton). Springer-Verlag, Berlin.

Gauthreaux, S. A. Jr. 1980. *Animal Migration, Orientation and Navigation*. Academic Press.

Gentle, M. J. 1975. Gustatory behaviour of the chicken and other birds. Pp. 305–318 in *Neural and Endocrine Aspects of Behaviour in Birds* (eds P. Wright, P. G. Caryl & D. M. Vowles). Elsevier, Amsterdam.

Gerritsen, A. F. C., van Heezik, Y. M. & Swennen, C. 1983. Chemoreception in two further *Calidris* species (*C. maritima* and *C. canutus*) with a comparison of the relative importance of chemoreception during foraging in *Calidris* species. *Netherl. J. Zool.* 33: 485–496.

Gibb, J. 1954. Feeding ecology of tits, with notes on three creeper and gold-crest. *Ibis* 96: 513–543.

Godfrey, W. E. 1966. *The Birds of Canada*. Queen's Printer, Ottawa.

Godfrey, W. E. 1967. Some winter aspects of the Great Gray Owl. *Can. Field Nat.* 81: 99–101.

Goodman, S. M. and Glynn, C. 1988. Comparative rates of natural osteological disorders in a collection of Paraguayan birds. *J. Zool (Lond.)* 214: 167–177.

Goodwin, D. 1986. *Crows of the World*. 2nd edition. British Museum (Natural History).

Gordon, I. E. 1989. *Theories of Visual Perception*. Wiley.

Goss-Custard, J. D. 1969. The winter feeding ecology of the Redshank *Tringa totanus*. *Ibis* 111: 338–356.

Goss-Custard, J. D. & Durell, S. E. A. le V. dit. 1987. Age-related effects in oystercatchers, *haematopus ostralegus*, feeding on mussels, *Mytilus edulis*. III. The effect of interference on overall intake rate. *J. Anim. Ecol.* 56: 549–558.

Gossard, E. E. & Hooke, W. H. 1975. *Waves in the Atmosphere*. Elsevier, Amsterdam.

Gottschaldt, K.-M. 1985. Structure and function of avian somatosensory receptors. Pp. 375–461 in *Form and Function in Birds*, vol. 3 (eds A. S. King & J. McLelland). Academic Press.

Gottschaldt, K.-M. & Lausmann, S. 1974. The peripheral morphological basis of tactile sensibility in the beak of geese. *Cell Tiss. Res.* 153: 477–496.

Goujon, E. 1869. Sur un appareil de corpuscles tactiles situe dans le bec des perroquets. *J. Anat. Physiol.* 6: 449–455.

Gould, J. L. 1982. The map sense of pigeons. *Nature* (Lond.) 296: 205–211.

Greenberg, R. 1983. The role of neophobia in determining the degree of foraging specialisation in some migrant warblers. *Am. Nat.* 122: 444–453.

Greenberg, R. 1987. Social facilitation does not reduce neophobia in chestnut-sided warblers (Parulinae: *Dendroica pensylvanica*). *J. Ethol.* 5: 7–10.

Gregory, R. L. 1974. *Concepts and Mechanisms of Perception*. Duckworth.

Griffin, D. R. 1954. Acoustic orientation in the oil bird, *Steatornis. Proc. Natl. Acad. Sci.* 39: 884–893.

Griffin, D. R. 1958. *Listening in the Dark*. Yale University Press, New Haven.

Griffin, D. R. 1973. Oriented bird migration in or between opaque cloud layers. *Proc. Am. Phil. Soc.* 117: 117–141.

Griffin, D. R., Hubbard, R. & Wald, G. 1947. The sensitivity of the human eye to infrared radiation. *J. Opt. Soc. America* 37: 546–552.

Griffin, D. R. & Suthers, R. A. 1970. Sensitivity of echolocation in cave swiftlets. *Biol. Bull.* 139: 495–501.

Griffin, D. R. & Thompson, D. 1982. Echolocation by cave swiftlets. *Behav. Ecol. Sociobiol.* 10: 119–123.

Grubb, T. C. 1972. Smell and foraging in shearwaters and petrels. *Nature* (Lond.) 237: 404–405.

Grubb, T. C. 1974. Olfactory navigation to the nesting burrow in Leach's Petrel (*Oceanodroma leucorrhoa*). *Anim. Behav.* 22: 192–202.

Grull, A. 1981. Investigations on the territory of the nightingale (*Luscinia megarhynchos*). *J. Orn.* 122: 259–284.

Gunter, R. 1951. The absolute threshold for vision in the cat. *J. Physiol. (Lond.)* 114: 8–15.

Gwinner, E. 1975. Circadian and circannual rhythms in birds. In *Avian Biology*, vol. 5 (eds D. S. Farmer & J. King). Academic Press.

Gwinner, E. 1985. Rhythms and time measurement. Pp. 510–513 in *A Dictionary of Birds* (eds B. Campbell & E. Lack). T. & A. D. Poyser.

Gwinner, E., Biebach, H. & von Kries, I. 1985. Food availability affects migratory restlessness in caged garden warblers (*Sylvia borin*). *Naturwissenschaften* 72: 51.

Haecker, V. 1916. Reizphysiologisches uber Vogelzug und Fruhgesang. *Biol. Zbl.* 36: 403–431.

Hailman, J. P. 1964. The Galapagos Swallow-tailed Gull is nocturnal. *Wilson Bull.* 76: 347–354.

Hald-Mortensen, P. 1970. Some preliminary notes from Tenerife. *Ibis* 112: 265–266.

Hale, W. D. 1980. *Waders*. Collins.

Hardy, A. R. 1977. Hunting ranges and feeding ecology of owls in farmland, Ph.D. thesis, Aberdeen University.

Harris, M. P. 1966. Breeding biology of the Manx Shearwater *Puffinus puffinus*. *Ibis* 108: 17–33.

Harris, M. P. 1970. Breeding ecology of the Swallow-tailed Gull *Creagrus furcatus*. *Auk* 87: 215–243.

Harris, R. I. 1970. Brownian motion in the cochlear partition. *J. Acoust. Soc. Am.* 44: 176–186.

Harrison, C. 1982. *An Atlas of the Birds of the Western Palaearctic*. Collins.

Harrison, C. J. O. 1978. *Bird Families of the World*. Elsevier Phaidon.

Harrison, P. 1983. *Seabirds: an identification guide*. Croom Helm, London.

Harrison, J. B. and Furumoto, L. 1971. Pigeon audiograms: comparison of evoked potential and behavioural thresholds in individual birds. *J. Aud. Res.* 11: 33–42.

Harrisson, T. 1976. The food of *Collacalia* swiftlets, (Aves, Apodidae) at Niah Great Cave in Borneo. *J. Bombay Nat. Hist. Soc.* 71: 376–393.

Hayman, P., Marchant, J. & Prater, T. 1986. *Shorebirds: an Identification Guide to the Waders of the World*. Croom Helm, London.

Hebrard, J. J. 1971. Fall nocturnal migration during two successive overcast days. *Condor* 74: 106–107.

Hecht, S., Hendley, C. D., Ross, S. & Richmond, P. M. 1948. The effect of exposure to sunlight on dark adaptation. *Am. J. Ophthalmol.* 31: 1573–1580.

Hecht, S. & Pirenne, M. H. 1940. The sensibility of the nocturnal long-eared owl in the spectrum. *J. Gen. Physiol.* 23: 709–717.

Henbest, N. 1989. Save our skies. *New Scientist* 121: 41–45.

Heppleston, P. B. 1971. The feeding ecology of oystercatchers *Haematopus ostralegus* L. in winter in Northern Scotland. *J. Anim. Ecol.* 41: 651–672.

Hilden, O. & Helo, P. 1981. The Great Grey Owl *Strix nebulosa* – a bird of the northern taiga. *Ornis Fennica* 58: 159–166.

Hilgerloh, G. 1989. Radar observations of passerine trans-Saharan migrants in Southern Portugal. *Ardeola* (in press).

Hills, B. L. 1980. Vision, visibility and perception in driving. *Perception* 9: 183–216.

Hinde, R. A. 1956. The biological significance of the territories of birds. *Ibis* 98: 340–369.

Hirons, G. J. M. 1976. A population study of the Tawny Owl (*Strix aluco*) and its main prey species in a woodland. D.Phil. thesis, Oxford University.

Hirons, G. J. M. 1985. The effects of territorial behaviour on the stability and dispersion of Tawny Owl (*Strix aluco*) populations. *J. Zool. Lond. (B)* 1: 21–48.

Hirons, G. J. M., Hardy, A. R. & Stanley, P. 1979. Starvation in young Tawny Owls. *Bird Study* 26: 59–63.

Hirsch, J. 1982. Falcon visual sensitivity to grating contrasts. *Nature* (Lond.) 300: 57–58.

Hocking, B. & Mitchell, B. L. 1961. Owl vision. *Ibis* 103: 281–284.

Hodos, W. & Leibowitz, R. W. 1977. Near-field acuity of pigeons: effects of scotopic adaptation and wavelength. *Vision Res.* 17: 463–467.

Hodos, W., Leibowitz, R. W. & Bonbright, J. C. 1976. Near-field acuity of pigeons: effects of head position and stimulus. *J. Exp. Anal. Behav.* 25: 129–141.

Hogan-Warburg, A. J. 1966. Social behaviour of the Ruff, *Philomachus pugnax* (L.) *Ardea* 54: 109–229.

Howard, E. 1920. *Territory in Bird Life*. John Murray, London.

Hughes, A. 1979. A schematic eye for the rat. *Vision Res.* 19: 569–588.

Hulscher, J. B. 1976. Location of cockles (*Cardium edule* L.) by the oystercatcher (*Haematopus ostralegus* L.) in darkness and daylight. *Ardea* 64: 292–310.

Hultsch, H. & Todt, D. 1981. Repertoire sharing and song-post distance in Nightingales (*Lusinia megarhynchos*). *Behav. Ecol. Sociobiol.* 8: 183–188.

Humphreys, G. W. & Bruce, V. 1988. *Visual Cognition.* Lawrence Erlbaum.

Hutchison, L. V. & Wenzel, B. M. 1980. Olfactory guidance in foraging by procellarii-forms. *Condor* 82: 314–319.

Iljitschew, W. D. 1974. Adaptionsokologische Parallelismenmosaikartige Evolution. Das Horsystem des Vogel als Objeckt der funktionellen Morphologie. *Biol. Zbl.* 93: 165–180.

Imber, M. J. 1973. The food of grey-faced petrels (*Pterodroma macroptera gouldi* (Hutton), with special reference to diurnal vertical migration of their prey. *J. Anim. Ecol.* 42: 645–662.

Jacobs, G. H., Crognale, M. & Fenwick, J. 1987. Cone pigment of the Great Horned Owl. *Condor* 89: 434–436.

James, P. C. 1986. How do Manx Shearwaters *Puffinus puffinus* find their burrows? *Ethology* 71: 287–294.

Jenkins, W. M. & Masterton, R. B. 1979. Sound localization in pigeon (*Columba livia*). *J. Comp. Physiol. Psychol.* 93: 403–413.

Johnson, C. S. 1986. Dolphin audition and echolocation capacities. In *Dolphin Cognition and Behaviour: a Comparative Approach* (eds R. J. Schusterman, J. Thomas & F. Wood). Lawrence Erlbaum.

Kacelnik, A. 1979. The foraging efficiency of Great Tits (*Parus major L.*) in relation to light intensity. *Anim. Behav.* 27: 237–241.

Kacelnik, A. and Krebs, J. F. 1982. The dawn chorus in the Great Tit (*Parus major*): proximate and ultimate causes. *Behaviour* 83: 287–309.

Kahl, M. P. & Peacock, L. J. 1963. The bill-snap reflex: a feeding mechanism in the American Wood Stork. *Nature* (Lond.) 199: 505–506.

Kajikawa, J. 1923. Beitrage zur Anatomie und Physiologie des Vogelauges. *Albrecht Graefe's Arch. Ophthalmol.* 112: 260–346.

Kallander, H. 1977. Piracy by Black-Headed Gulls on Lapwings. *Bird Study* 24: 186–194.

Kamil, A. C. & Balda, R. P. 1985. Cache recovery and spatial memory in Clark's nutcracker (*Nucifraga columbiana*). *J. Exptl. Psychol: Anim. Behav. Processes* 11: 95–111.

Kare, M. R. & Rogers, J. G. 1976. Sense organs. Taste. In *Avian Physiology* (ed. P. D. Sturkie). Springer-Verlag, Berlin.

Kear, J. 1960. Food selection in certain finches with special reference to interspecific differences. Ph.D. thesis, Cambridge University (quoted in Berkhoudt, H. 1985).

Keeton, W. T. 1979a. Avian orientation and navigation: a brief overview. *British Birds* 72: 451–470.

Keeton, W. T. 1979b. Avian orientation and navigation. *Ann. Rev. Physiol.* 41: 353–366.

Kellogg, W. N. 1962. Sonar system of the blind. *Science* 137: 399–404.

Kemp. A. & Calburn, S. 1987. *The Owls of Southern Africa.* New Holland.

Kerlinger, P. & Gauthreaux, S. A. Jr. 1985a. Seasonal timing, geographical distribution, and flight behaviour of broad-winged hawks during spring migration in Texas. A radar and visual study. *Auk* 102: 735–743.

Kerlinger, P. & Gauthreaux, S. A. Jr. 1985b. Flight behaviour of raptors during spring migration in south Texas studied with radar and visual observation. *J. Field Ornithol.* 56: 394–402.

Kerlinger, P. & Moore, F. R. 1989. Atmospheric structure and avian migration. In *Current Ornithology Vol. 6* (ed. D. Power). Plenum Press, New York.

King, A. S. & King, D. Z. 1980. Avian morphology: general principles. Pp. 1–38 in *Form and Function in Birds,* vol. 1 (eds A. S. King & J. McLelland). Academic Press.

Klump, G. M., Windt, W. & Curio, E. 1986. The great tit's (*Parus major*) auditory resolution in azimuth. *J. Comp. Physiol. A.* 158: 383–390.

Knudsen, E. I. 1980. Sound localisation in birds. Pp. 289–322 in *Comparative Studies of Hearing in Vertebrates* (eds A. N. Popper & R. R. Fay). Springer-Verlag, Berlin.

Knudsen, E. I. & Konishi, M. 1979. Mechanisms of sound localisation in the barn owl (*Tyto alba*). *J. Comp. Physiol.* 133: 13–21.

Konishi, M. 1973a. How the owl tracks its prey. *Am. Sci.* 61: 414–424.

Konishi, M. 1973b. Locatable and non-locatable acoustic signals for barn owls. *Am. Nat.* 107: 775–785.

Korpimaki, E. 1981. On the ecology and biology of Tengmalm's Owl (*Aegolius funereus*). *Acta Universitatis Ouluensis*, series A, 118.

Krantz, P. E. & Gauthreaux, S. A. Jr. 1975. Solar radiation, light intensity and roosting behaviour in birds. *Wilson Bull.* 87: 91–95.

Krebs, J. R. 1971. Territory and breeding density in the Great Tit *Parus major* L. *Ecology* 52: 2–22.

Kuhne, R. & Lewis, B. 1985. External and middle ears. Pp. 227–271 in *Form and Function in Birds,* vol. 3 (eds A. S. King & J. McLelland). Academic Press.

Lack, D. 1956. *Swifts in a Tower.* Methuen.

Lack, D. 1960. The height of bird migration. *British Birds* 53: 5–10.

Lack, D. 1965. *The Life of the Robin.* Collins.

Lange, G. 1968. Uber Nahrung, Nahrungsaufnahme und Verdaugstrakt mitteleuropaischer Limikolen. *Beitr. Vogelkunds* 13: 225–234.

Langham, N. 1980. Breeding biology of the Edible-nest swiftlet *Aerodramus fuciphagus. Ibis* 122: 447–461.

Leask, M. J. M. 1977. A physicochemical mechanism for magnetic field detection by migratory birds and homing pigeons. *Nature* 267: 144–145.

Le Grand, Y. 1957. *Light, Colour and Vision.* Translated by R. Hunt, J. Walsh and F. Hunt. Chapman & Hall.

Leopold, A. & Eynon, A. E. 1961. Avian daybreak and morning song in relation to time and light intensity. *Condor* 63: 269–293.

Lindblad, J. 1967. *I Ugglemarker.* Bonniers, Stockholm.

Lundberg, A. 1979. Residency, migration and a compromise: adaptations to nest site scarcity and food specialization in three Fenoscandian owl species. *Oecologia* (Berl.) 41: 273–281.

Lutz, F. E. 1931. Light as a factor controlling the start of daily activity of a wren and stingless bees. *Am. Mus. Novit.* 468: 1–9.

Lythgoe, J. N. 1979. *The Ecology of Vision.* Oxford University Press.

Macdonald, D. W. 1976. Nocturnal observations of Tawny Owls *Strix aluco* preying upon earthworms. *Ibis* 118: 579–580.

Macdonald, D. W. 1987. *Running with the Fox.* Unwin Hyman.

Mace, R. 1986. Importance of female behaviour in the dawn chorus. *Animal Behav.* 34: 621–622.

Mace, R. 1987a. Why do birds sing at dawn. *Ardea* 75: 123–132.

Mace, R. 1987b. The dawn chorus in the great tit *Parus major* is directly related to female fertility. *Nature (Lond.)* 330: 745–746.

Mace, R. 1989. The relationship between daily routines of singing and foraging: an experiment on captive Great Tits *Parus major. Ibis* 131: 415–420.

McFadden, S. A. & Reymond, L. 1985. A further look at the binocular visual field of the pigeon (*Columba livia*). *Vision Res.* 25: 1741–1746.

Macphail, E. M. 1986. Animal memory: past, present and future. *Quart. J. Exptl. Psychol.* 38B: 349–364.

Manley, G. A. 1972. A review of some current concepts of the functional evolution of the ear in terrestrial vertebrates. *Evolution* 26: 608–621.

Marti, C. D. 1974. Feeding ecology of four sympatric owls. *Condor* 76: 45–61.

Martin, G. R. 1974. Color vision in the tawny owls *Strix aluco. J. Comp. Physiol. Psychol.* 86: 133–141.

Martin, G. R. 1977. Absolute visual threshold and scotopic spectral sensitivity in the Tawny Owl, *Strix aluco. Nature* (Lond.) 268: 636–638.

Martin, G. R. 1982. An owl's eye: schematic optics and visual performance in *Strix aluco. L. J. Comp. Physiol.* 145: 341–349.

Martin, G. R. 1983. Schematic eye models in vertebrates. Pp. 43–81 in *Progress in Sensory Physiology,* vol. 4 (ed. D. Ottoson). Springer-Verlag, Berlin.

Martin, G. R. 1984a. The visual fields of the Tawny Owl, *Strix aluco L. Vision Res.* 24: 1739–1751.

Martin, G. R. 1984b. An environmental limit on absolute auditory sensitivity in non-aquatic vertebrates. *Behav. Processes* 9: 205–221.

Martin, G. R. 1986a. Sensory capacities and the nocturnal habit in owls. *Ibis* 128: 266–277.

Martin, G. R. 1986b. Total panoramic vision in the mallard duck (*Anas platyrhynchos*). *Vision Res.* 26: 1303–1305.

Martin, G. R. 1986c. The eye of a Passeriforme bird, the European Starling (*Sturnus vulgaris*): eye movement amplitude, visual fields and schematic optics. *J. Comp. Physiol.* 159: 545–557.

Martin, G. R. 1990. The visual problems of nocturnal migration. Pp. 186–197 in *Bird migration: the physiology and ecophysiology* (ed. E. Gwinner). Springer-Verlag.

Martin, G. R. & Gordon, I. E. 1974a. Increment-threshold spectral sensitivity in the Tawny Owl (*Strix aluco*). *Vision Res.* 14: 615–620.

Martin, G. R. & Gordon, I. E. 1974b. Visual acuity in the Tawny Owl (*Strix aluco*). *Vision Res.* 14: 1393–1397.

Martin, G. R., Gordon, I. E. & Cadle, D. R. 1975. Electroretinographically determined spectral sensitivity in the Tawny Owl (*Strix aluco*). *J. Comp. Physiol. Psychol.* 89: 72–78.

Martin, G. R. & Young, S. R. 1983. The retinal binocular field of the pigeon (*Columba livia*): English racing homer. *Vision Res.* 23: 911–915.

Martinoya, C., Rey, J. & Bloch, S. 1981. Limits of the pigeon's binocular field and direction for best binocular viewing. *Vision Res.* 21: 1197–1200.

Masterton, B., Heffner, H. & Ravizza, R. 1969. The evolution of human hearing. *J. Acoust. Soc. Am.* 45: 966–985.

Matthews, L. H. & Matthews, B. H. C. 1939. Owls and infra-red radiation. *Nature* (Lond.) 143: 983.

Mattison, C. 1986. *Snakes of the World.* Blandford Press.

McNamara, J. M., Mace, R. H. and Houston, A. I. 1987. Optimal daily routines of singing and foraging in a bird singing to attract a mate. *Behav. Ecol. Sociobiol.* 20: 399–405.

Mead, C. 1983. *Bird Migration.* Country Life.

Medway, Lord. 1959. Echolocation among *Collocalia. Nature* (Lond.) 184: 1352–1353.

Medway, Lord. 1962a. The swiftlets (*Collocalia*) of Niah cave, Sarawak, part 1. *Ibis* 104: 45–66.

Medway, Lord. 1962b. The swiftlets (*Collocalia*) of Niah cave, Sarawak, part 2. *Ibis* 104: 228–245.

Medway, Lord & Pye, J. D. 1977. Echolocation and the systematics of swiftlets. Pp. 225–238 in *Evolutionary Ecology* (eds B. Stonehouse & C. Perrins). Macmillan.

Merton, D. V., Morris, R. B. & Atkinson, I. A. E. 1984. Lek behaviour in a parrot: the Kakapo *Strigops habroptilus* of New Zealand. *Ibis* 126: 277–283.

Middleton, W. E. K. 1958. *Vision Through the Atmosphere.* University of Toronto Press, Toronto.

Mikkola, H. 1983. *Owls of Europe.* T. & A. D. Poyser.

Mills, A. W. 1958. On the minimum audible angle. *J. Acoust. Soc. Am.* 30: 237–246.

Mooney, N. 1982. Piracy in falcons. *Australas. Raptor Ass. News* 3: 13.

Moore, F. R. 1987. Sunset and the orientation behaviour of migrating birds. *Biol. Rev.,* 62: 65–86.

Morris, D. 1955. The seed preferences of certain finches under controlled conditions. *Avic. Mag.* 61: 271–287.

Muller-Schwarze, D. & Mozell, M. (eds). 1977. *Chemical Signals in Vertebrates.* Plenum Press.

Nachmias, J. 1972. Signal detection theory and its application to problems in vision. Pp. 56–77 in *Handbook of Sensory Physiology*, vol. VII/4 (eds D. Jameson and L. M. Hurvich). Springer-Verlag, Berlin.

National Geographic Society. 1983. *Field Guide to the Birds of North America.* National Geographic Society, Washington.

Natural Illumination Charts. 1952. US Navy Research and Development project NS 714–100. Report No. 374–1 (September).

Nelson, J. B. 1968. Breeding behaviour of the Swallow-tailed Gull in the Galapagos. *Behaviour* 30: 146–174.

Nero, R. 1980. *The Great Gray Owl: Phantom of the Northern Forest.* Smithsonian Inst. Press, Washington.

Nettleship, D. N. & Birkhead, T. R. 1986. *The Atlantic Alcidae.* Academic Press.

Newman, E. A. & Hartline, P. H. 1981. Integration of visual and infrared information in bimodal neurons of the Rattlesnake optic tectum. *Science* 213: 789–791.

Newton, I. 1979. *Population Ecology of Raptors.* T. & A. D. Poyser.

Newton, I. 1985. Hawk. P. 276 in *A Dictionary of Birds* (eds B. Campbell & E. Lack). T. & A. D. Poyser.

Neyrolles, J.-N. 1985. Kagu. P. 314 in *A Dictionary of Birds* (eds B. Campbell & E. Lack). T. & A. D. Poyser.

Nicol, J. A. C. & Arnott, H. J. 1974. Tapeta lucida in the eyes of goatsuckers (*Caprimulgidae*). *Proc. R. Soc. B.* 187: 349–352.

Nisbet, I. C. T. 1970. Autumn migration of the blackpoll warbler: evidence for long flight provided by regional survey. *Bird Banding* 41: 207–240.

Norberg, R. A. 1968. Physical factors in directional hearing in *Aegolius funereus* (Stringiformes), with special reference to the significance of the asymmetry of the external ears. *Ark. Zool.* 20: 181–204.

Norberg, R. A. 1977. Occurrence and independent evolution of bilateral ear asymmetry in owls and implications on owl taxonomy. *Phil. Trans. R. Soc.* 280: 376–408.

Norberg, R. A. 1978. Skull asymmetry, ear structure and function and auditory localization in Tengmalm's Owl, *Aegolius funereus L. Phil. Trans. R. Soc.* 282B: 325–410.

Novick, A. 1959. Acoustic orientation in the cave swiftlet. *Biol. Bull.* 117: 497–503.

Ogilvie, M. 1976. *The Winter Birds.* Michael Joseph.

Olton, D. S., Handelmann, G. E. & Walker, J. A. 1981. Spatial memory and food searching strategies. In: Kamil, A. C. & Sargent, T. D. (eds) *Foraging Behaviour,* pp. 333–354, New York, Garland Press.

Ostmann, O. W., Ringer, R. K. & Tetzlaff, M. 1963. The anatomy of the feather follicle and its immediate surroundings. *Poult. Sci.* 42: 958–969.

Owen, D. F. 1954. The winter weight of titmice. *Ibis* 96: 299–309.

Palmgren, P. 1944. Studien uber die Tagesrhythmik gekafigter Zugvogel. *Z. Tierpsy-chol.* 6: 44–86.

Papi, F. 1982. Olfaction and homing in pigeons: ten years of experiments. Pp. 149–159 in *Avian Navigation* (eds F. Papi & H. G. Wallraff). Springer-Verlag, Berlin.

Papi, F. & Wallraff, H. G. (eds) 1982. *Avian Navigation.* Springer-Verlag, Berlin.

Parslow, J. L. F. 1969. The migration of passerine night migrants studied by radar. *Ibis* 111: 48–79.

Payne, R. S. 1962. How the Barn Owl locates prey by hearing. *Living Bird* 1: 151–159.

Payne, R. S. 1971. Acoustic location of prey by Barn Owls (*Tyto alba*). *J. Exp. Biol.* 54: 535–573.

Payne, R. S. & Drury, W. H. 1958. Marksman of the darkness. *Nat. Hist. (N.Y.)* 67: 316–323.

Pennycuick, C. J. 1960. The physical basis of astronavigation in birds: theoretical considerations. *J. Exp. Biol.* 37: 573–593.

Pennycuick, C. J. 1969. The mechanics of bird migration. *Ibis* 111: 48–79.

Perrins, C. M. 1979. *British Tits.* Collins.

Peters, J. L. 1931. *Check-list of Birds of the World,* vol. 1. Harvard University Press, Cambridge, Mass.

Pettigrew, J. D. & Frost, B. J. 1985. A tactile fovea in the *Scolopacidae? Brain Behav. Evol.* 26: 185–195.

Pettigrew, J. D. and Konishi, M. 1984. Some observations in the visual system of the Oilbird *Steatornis caripensis.* National Geographical Society Research Reports 16: 439–449.

Pienkowski, M. W. 1982. Diet and energy intake of Grey and Ringed Plovers, *Pluvialis squatarola* and *Charadrius hiaticula,* in the non-breeding season. *J. Zool.* 197: 511–549.

Pirenne, M. H. 1962. Absolute thresholds and quantum effects. Pp. 123–140 in *The Eye,* vol. 2 (ed. H. Davson). Academic Press.

Pirenne, M. H., Marriott, F. H. C. & O'Doherty, E. F. 1957. Individual differences in night vision efficiency. *Med. Res. Counc. G.B. Spec. Rep. Ser.* 294.

Poulter, T. C. 1969. Sonar of penguins and fur seals. *Proc. Calif. Acad. Sci.* 36: 363–380.

Presti, D. E. 1985. Avian navigation, geomagnetic field sensitivity, and biogenic magnetite. Pp. 455–482 in *Magnetite Biomineralization and Magnetoreception in Organisms* (eds J. L. Kirschvink, S. Jones & B. J. MacFadden). Plenum Press, New York.

Price-Jones, D. 1983. *Australian Birds of Prey.* Doubleday, Australia.

Prince, P. A. & Francis, M. D. 1984. Activity budgets of foraging Grey-headed Albatrosses. *Condor* 86: 297–300.

Pumphrey, R. J. 1948. The sense organs of birds. *Ibis* 90: 171–199.

Pye, J. D. 1979. Why ultrasound? *Endeavour* n.s. 3: 57–62.

Pye, J. D. 1980. Echolocation signals and echoes in air. Pp. 309–353 in *Animal Sonar Systems* (eds R.-G. Busnel and J. F. Fish). Plenum Press, New York.

Pye, J. D. 1985. Echolocation. Pp. 165–166 in *A Dictionary of Birds* (eds B. Campbell & E. Lack). T. & A. D. Poyser.

Ranft, R. & Slater, P. J. B. 1987. Absence of ultrasonic calls from night-flying Storm Petrels *Hydrobates pelagicus. Bird Study* 34: 92–93.

Rankin, M. A. (ed.) 1985. *Migration: Mechanisms and Adaptive Significance.* University of Texas, Houston.

Ratcliffe, D. A. 1980. *The Peregrine Falcon.* T. & A. D. Poyser.

Ravenscroft, N. O. M. 1989. The status and habitat of the Nightjar *Caprimulgus europaeus* in coastal Suffolk. *Bird Study* 36: 161–169.

Rayner, J. M. V. 1985. Flight, speeds of. Pp. 224–226 in *A Dictionary of Birds* (eds B. Campbell & E. Lack). T. & A. D. Poyser.

Reid, B. & Williams, G. R. 1975. The Kiwi. Pp. 301–330 in *Biogeography and Ecology in New Zealand* (ed. G. Kuschel). Junk, The Hague.

Reymond, L. 1987. Spatial visual acuity of the falcon, *Falco berigora*: a behavioural, optical and anatomical investigation. *Vision Res.* 27: 1859–1874.

Richardson, W. J. 1972. Autumn migration and weather in eastern Canada: a radar study. *Am. Birds* 26: 10–17.

Richardson, W. J. 1976. Autumn migration over Puerto Rico and the Western Atlantic: a radar study. *Ibis* 118: 309–332.

Richardson, W. J. 1978. Reorientation of nocturnal landbird migrants over the Atlantic Ocean near Nova Scotia in autumn. *Auk* 95: 717–732.

Richardson, W. J. 1979. Radar techniques for wildlife studies. *Nat. Wildl. Fed. Sci. Tec. Ser.* 3: 171–179.

Riggs, L. A. 1965. Light as a stimulus for vision. Pp. 1–38 in *Vision and Visual Perception* (ed. C. H. Graham). Wiley, New York.

Rijnsdoorp, A., Daan, S. & Djkstra, C. 1981. Hunting in the Kestrel *Falco tinnunculus* and the adaptive significance of daily habits. *Oecolgia (Berl.)* 50: 391–406.

Robert, M. & McNeil, R. 1989. Comparative day and night feeding strategies of shore bird species in a tropical environment. *Ibis* 131: 69–79.

Robert, M., McNeil, R. & Leduc, A. 1989. Conditions and significance of night feeding in shorebirds and other water birds in a tropical lagoon. *Auk* 106: 94–101.

Rochon-Duvigneaud, A. 1943. *Les Yeux et la Vision de Vertebres.* Masson, Paris.

Rodieck, R. W. 1973. *The Vertebrate Retina: Principles of Structure and Function.* Freeman, San Francisco.

Roffler, S. K. & Butler, R. A. 1968. Factors that influence the localization of sound in the vertical plane. *J. Acoust. Soc. Am.* 43: 1255–1259.

Sales, G. D. & Pye, J. D. 1974. *Ultrasonic Communication by Animals.* Chapman & Hall.

Salomonsen, F. 1950. *Gronlands fugle. The Birds of Greenland.* Munksgaard, Copenhagen.

Sandberg, R., Pettersson, J. & Alerstam, T. 1988. Shifted magnetic fields lead to deflected and axial orientation of migrating robins, *Erithacus rubecula,* at sunset. *Anim. Behav.* 36: 877–887.

Sandel, T. T., Teas, D. C., Feddersen, W. E. & Jeffress, L. A. 1955. Localisation of sound from single and paired sources. *J. Acoust, Soc. Am.* 27: 842–852.

Sauer, E. G. F. & Sauer, E. M. 1955. Zur Frage der nächtlichen Zugorientierung von Grasmucken. *Rev. Suisse Zool.* 62: 250–259.

Schlegel, R. 1969. *Der Ziegenmelker.* Wittenberg, Lutherstadt.

Schmidt, P. H., van Gemert, A. H., Der Vries, R. J. & Duyff, J. W. 1953. Binaural thresholds for azimuth difference. *Acta. Physiol. et Pharmacol. Neer.* 3: 2–18.

Schmidt-Koenig, K. & Keeton, W. T. (eds) 1978. *Animal Migration, Navigation and Homing.* Springer-Verlag, Berlin.

Schmitz, J. & Middel, A. 1966. Die Abhangigkeit des Vogelgesanges von der Helligkeit. *Orn. Mitt.* 18: 111–114.

Schodde, R. & Mason, I. J. 1980. *Nocturnal Birds of Australia.* Melbourne.

Schwan, A. 1920. Vogelgesang und wetter, physikalisch-biologisch Untersucht. Vorlaufige Mitteilung. *Pflug. Arch. ges. Physiol.* 180: 341–347.

Schwartzkopff, J. 1962. Zur Frage der Richtungshorens von Eulen (Striges). *Z. Vergl. Physiol.* 45: 570–580.

Sears, H. F., Moseley, L. J. & Mueller, H. C. 1976. Behavioural evidence on skimmers' evolutionary relationships. *Auk* 93: 170–174.

Semm, P., Nohr, D., Demaine, C. & Wiltschko, W. 1984. Neural basis of magnetic compass: interactions of visual, magnetic and vestibular inputs in the pigeon's brain. *J. Comp. Physiol. A.* 155: 283–288.

Semm, P. & Demaine, C. 1986. Neurophysiological properties of magnetic cells in the pigeon's visual system. *J. Comp. Physiol. A.* 159: 619–625.

Serventy, D. L. 1936. Feeding methods of Podargus. *Emu* 36: 74–90.

Serventy, D. L. 1985a. Frogmouth. P. 244 in *A Dictionary of Birds* (eds B. Campbell & E. Lack). T. & A. D. Poyser.

Serventy, D. L. 1985b. Owlet-frogmouth. Pp. 421–422 in *A Dictionary of Birds* (eds B. Campbell & E. Lack). T. & A. D. Poyser.

Shallenberger, R. J. 1975. Olfactory use in the wedge-tailed shearwater (*Puffinus pacificus*) on Manana Island, Hawaii. In *Olfaction and Taste* (eds D. A. Denton & J. P. Coghlan). Academic Press.

Sheldon, W. G. 1967. *The Book of the American Woodcock*. Amherst.

Sheppard, M. B. and Spitzer, P. R. 1985. Feeding behaviour and social ecology of the Steward Island Kakapo. National Geographic Society Research Report 20: 657–679.

Shlaer, S. 1937. The relation between visual acuity and illumination. *J. Gen. Physiol.* 21: 165–188.

Sibley, C. G. & Ahlquist, J. E. 1972. A comparative study of egg white proteins of non-passerine birds. *Bull. Peabody Mus. Nat. Hist.* 39: 1–276.

Sibley, C. G., Ahlquist, J. E. and Monroe, B. L. 1988. A classification of the living birds of the world based upon DNA–DNA hybridization studies. *Auk.* 105: 409–423.

Simmons, J. A. 1969. Acoustic radiation patterns for the echolocating bats *Chilonycteris rubiginosa* and *Eptesicus fuscus. J. Acoust. Soc. Am.* 46: 1054–1056.

Simmons, J. A., Howell, D. S. & Suga, N. 1975. The information content of bat sonar sound. *Am. Scientist* 63: 204–215.

Sivak, J. G. & Howland, H. C. 1987. Refractive state of the eye of the brown kiwi (*Apteryx australis*). *Can. J. Zool.* 65: 2833–2835.

Sivian, L. J. & White, S. D. 1933. On minimal audible sound fields. *J. Acoust. Soc. Am.* 4: 288–321.

Skutch, A. F. 1970. Life history of the Common Potoo. *Living Bird* 9: 265–280.

Smith, C. A. 1981. Recent advances in structural correlates of auditory receptors. In *Progress in Sensory Physiology,* vol. 2 (ed. D. Ottoson). Springer-Verlag, Berlin.

Smith, C. A. 1985. Inner ear. Pp. 273–310 in *Form and Function in Birds,* vol. 3 (eds A. S. King & J. McLelland). Academic Press.

Smythe, D. M. & Roberts, J. R. 1983. The sensitivity of echolocation by the Grey Swiftlet *Aerodramus spodiopygius. Ibis* 125: 339–345.

Snow, B. W. & Snow, D. W. 1968. Behaviour of the Swallow-tailed Gull of the Galapagos. *Condor* 70: 252–264.

Snow, D. W. 1953. The migration of the Greenland Wheateater. *Ibis* 95: 376–378.

Snow, D. W. 1961. The natural history of the Oilbird, *Steatornis caripensis,* in Trinidad. 1. General behaviour and breeding habits. *Zoologica* 46: 27–48.

Snow, D. W. 1962. The natural history of the Oilbird, *Steatornis caripensis,* in Trinidad. 2. Population, breeding ecology and food. *Zoologica* 47: 199–221.

Snow, D. W. 1976. *The Web of Adaptation: Bird Studies in the American Tropics.* Collins.

Snow, D. W. 1988. *A Study of Blackbirds.* 2nd edition. British Museum.

Snow, D. W. & Nelson, J. B. 1984. Evolution and adaptations of Galapagos sea-birds. *Biological J. Linn. Soc.* 21: 137–155.

Snyder, A. W., Laughlin, S. B. & Stavenga, D. G. 1977. Information capacity of eyes. *Vision Res.* 17: 1163–1175.

Snyder, D. E. 1957. *Arctic Birds of Canada.* Toronto.

Southern, H. N. 1970. The natural control of a population of Tawny Owls (*Strix aluco*). *J. Zool. Lond.* 162: 197–285.

Southern, H. N. & Lowe, V. P. 1968. The pattern and distribution of prey and predation in Tawny owl territories. *J. Anim. Ecol.* 37: 75–97.

Sparks, J. & Soper, T. 1970. *Owls: their Natural and Unnatural History.* David & Charles.

Spencer, K. G. 1953. *The Lapwing in Britain.* A. Brown and Son.

Stattelman, A. J., Talbot, R. B. & Coulter, D. B. 1975. Olfactory thresholds of pigeons (*Columba livia*), quail (*Colinus virginianus*) and chickens (*Gallus gallus*). *Comp. Biochem. Physiol.* 50A: 807–809.

Stebbins, W. C. (ed.) 1970. *Animal Psychophysics.* Appleton-Century-Crofts, New York.

Steinbach, M. J. & Money, K. E. 1973. Eye movements of the owl. *Vision Res.* 13: 889–891.

Stribling, H. L. & Doerr, P. D. 1985. Nocturnal use of fields by American Woodcock. *J. Widl. Mgmt.* 49: 485–491.

Suthers, R. A. & Hector, D. W. 1985. The physiology of vocalizations by the echolocating Oilbird, *Steatornis caripensis. J. Comp. Physiol. A.* 156: 243–266.

Tannabaum, B. and Wrege, P. H. 1984. Radiotelemetry study of the oilbird. National Geographical Society Research Report, 17: 843–854.

Tansley, K. 1965. *Vision in Vertebrates.* Chapman & Hall.

Tomback, D. F. 1980. How nutcrackers find their seed stores. *Condor* 82: 10–19.

Tyron, C. A. 1943. The Great Gray Owl as a predator on pocket gophers. *Wilson Bull.* 55: 130–131.

Ullman, S. 1980. Against direct perception. *The Behavioral and Brain Sciences, 3* (whole issue).

Vanderplanck, F. L. 1934. The effect of infra-red waves on Tawny Owls (*Strix aluco*). *J. Zool. Lond.* 505–507.

Vander Wall, S. B. 1982. An experimental analysis of cache recovery in Clark's nutcracker. *Anim. Behav.* 30: 84–94.

Van Dijk, T. 1973. A comparative study of hearing in owls of the family Strigidae. *Nethl. J. Zool.* 23: 131–167.

Van Heezik, Y. M., Gerritsen, A. E. C. & Swennen, C. 1983. The influence of chemoreception on the foraging behaviour of two species of sandpiper, *Calidris alba* and *Calidris alpina. Netherl. J. Sea Res.* 17: 47–56.

Van Rhijn, J. G. 1983. On the maintenance and origin of alternative strategies in the Ruff *Philomachus pugnax. Ibis* 125: 482–498.

Vaughan, W. & Greene, S. L. 1984. Pigeon visual memory capacity. *J. Exptl. Psychol.: Anim. Behav. Processes* 10: 256–271.

Verheijen, F. J. 1980. The moon: a neglected factor in studies on the collision of nocturnal migrant birds with tall lighted structures and with aircraft. *Die Vogelwarte* 30: 305–320.

Verheijen, F. J. 1981. Bird kills at lighted man-made structures: not on nights close to full moon. *Am. Birds* 35: 251–254.

Voous, K. H. 1985. Table of classification. Pp. xi–xvii in *A Dictionary of Birds* (eds B. Campbell & E. Lack). T. & A. D. Poyser.

Voous, K. H. 1988. *Owls of the Northern Hemisphere.* Collins.

Walcott, B. & Walcott, C. 1982. A search for magnetic field receptors in animals. Pp. 338–343 in *Avian Navigation* (eds F. Papi & H. G. Wallraff). Springer-Verlag, Berlin.

Walls, G. L. 1942. *The Vertebrate Eye and its Adaptive Radiation.* Cranbrook Institute of Science, Michigan.

Walsh, J. W. T. 1958. *Photometry.* Constable.

Walter, H. 1979. Eleonora's Falcon: *Adaptation to Prey and Habitat in a Social Raptor.* University of Chicago, Chicago.

Walters, M. P. 1978. Kagu. Pp. 88–89 in *Bird Families of the World* (ed. C. J. O. Harrison). Elsevier-Phaidon.

Wardough, A. A. 1984. Wintering strategies of British Owls. *Bird Study* 31: 76–77.

Warham, J. 1955. Observations of the little Shearwater at the nest. *West. Aus. Nat.* 5: 31–39.

Warham, J. 1958. The nesting of the Shearwater *Puffinus carneipes. Auk* 75: 1–13.

Warham, J. 1960. Some aspects of breeding behaviour in the short-tailed shearwater. *Emu* 60: 75–87.

Watson, D. 1977. *The Hen Harrier.* T. & A. D. Poyser.

Weale, R. A. 1960. *The Eye and its Function.* Hatton Press.

Wendland, V. 1984. The influence of prey fluctuations on the breeding success of the Tawny Owl (*Strix aluco*). *Ibis* 126: 284–295.

Wenzel, B. M. 1968. Olfactory prowess of the Kiwi. *Nature* (Lond.) 220: 1133–1134.

Wever, E. G. & Lawrence, M. 1954. *Physiological Acoustics.* Princeton University Press, Princeton.

Whiten, A. 1978. Operant studies on pigeon orientation and navigation. *Anim. Behav.* 26: 571–610.

Wilson, E. O. 1975. *Sociobiology.* Harvard University Press, Cambridge, Mass.

Wiltschko, W. & Wiltschko, R. 1975a. The interaction of stars and magnetic field in the orientation system of night migrating birds. 1. Autumn experiments with European warblers (Gen *Sylvia*). *Z. Tierpsychol.* 37: 337–355.

Wiltschko, W. & Wiltschko, R. 1975b. The interaction of stars and magnetic field in the orientation system of night migrating birds. 2. Spring experiments with European robins (*Erithacus rubecula*). *Z.Tierpsychol.* 39: 265–282.

Wiltschko, W. & Wiltschko, R. 1988a. Migratory orientation: magnetic and celestial factors work together. *J. Ornithol.* 129: 265–286.

Wiltschko, W. & Wiltschko, R. 1988b. Magnetic versus celestial orientation in migrating birds. *Trends in Evolution and Ecology* 3: 13–15.

Wink, M., Wink, C. & Ristow, D. 1980. *Konnen Sich Gelbschnabelsturmtaucher durch Echolotung Orientieren?* Vortrag DOG-Tagung, Hanover.

Wolf, L. L., Stiles, F. G. & Hainsworth, F. R. 1976. Ecological organization of a tropical highland hummingbird community. *J. Anim. Ecol.* 45: 349–379.

Wood, B. 1982. The trans-Saharan spring migration of Yellow Wagtails (*Motacilla flava*). *J. Zool. Lond.* 197: 267–283.

Wood, C. A. 1917. *The Fundus Occuli of Birds Especially as Viewed by the Ophthalmoscope.* Lakeside Press, Chicago.

Wright, A. A. 1979. Color-vision psychophysics: a comparison of pigeon and human. Pp. 89–127 in *Neural Mechanisms of Behaviour in the Pigeon* (eds A. Granda & J. H. Maxwell). Plenum Press.

Wyszecki, G. & Stiles, W. S. 1967. *Color Science.* John Wiley, New York.

Zeigler, H. P., Levitt, P. W. & Levine, R. R. 1980. Eating in the pigeon (*Columba livia*): movement patterns, stereotypy and stimulus control. *J. Comp. Physiol. Psychol.* 94: 783–794.

Zusi, R. L. 1962. Structural adaptations of the head and neck of the Black Skimmer, *Rynchops nigra*. *Publ. Nuttall Orn. Club* 3: 1–101.

Zusi, R. L. 1985. Skimmer. Pp. 546–547 in *A Dictionary of Birds* (eds B. Campbell & E. Lack). T. & A. D. Poyser.

Zusi, R. L. and Bridge, D. 1981. On the slit pupil of the Black Skimmer (*Rynchops niger*). *J. Field. Ornith.* 52: 338–340.

Zweers, G. A. 1982a. The feeding system of the pigeon (*Columba livia L.*). *Adv. Anat. Embryol. Cell Biol.* 73: 1–108.

Zweers, G. A. 1982b. Pecking in the pigeon (*Columba livia*) L. *Behaviour* 81: 173–230.

Zweers, G. A. & Wouterlood, F. G. 1973. Functional anatomy of the feeding apparatus of the mallard (*Anas platyrhynchos*). *Proc. 3rd Eur. Anat. Congr. Manchester* 88–89.

Index

Page numbers in italic refer to major discussions in the text

224   *Index*